# A STORY ABOUT THE VOID.
# AND THE SEA. AND PHYSICS. AND LOVE. AND THE STONES.

Jaitzul

Copyright © 2024 Jaitzul

All rights reserved

The characters and events portrayed in this book are fictitious. Any similarity to real persons, living or dead, is coincidental and not intended by the author.

No part of this book may be reproduced, or stored in a retrieval system, or transmitted in any form or by any means, electronic, mechanical, photocopying, recording, or otherwise, without express written permission of the publisher.

To all the people who have not been able to fully realize their potential.
And to all those who have suffered the consequences of doing so.
To you, the one reading these words.
And to all the people who illuminate me from the past.

~Jaitzul~, 2024.

# CONTENTS

Title Page
Copyright
Dedication
Memories of Sestao     1
The void.     5
And the sea.     9
Memories of Murcia.     11
Thunder and lightning.     13
Sharing the present.     16
Disordering Time.     18
Tearing apart the sea.     22
Memories of Oriñón.     24
Ch-ch-ch-changes.     27
Cosmos.     29
The arrow of time.     31
Disordering Time to Order the World.     33
And Physics.     35
Newton was ugly.     37
Tearing The World Apart.     40
Disordering the World.     43
Ordered Elements.     45

| | |
|---|---|
| Butterfly Effect. | 48 |
| Schrödinger's Cat. | 51 |
| The Observer. | 54 |
| Disordering Time II. | 57 |
| Memories of Santiago. | 59 |
| Singularity. | 62 |
| Cupid's Arrow. | 64 |
| And love. | 65 |
| Memories of the Big Bang. | 67 |
| Quantum Superposition. | 71 |
| Everything is Relative. | 75 |
| Memories of Galway. | 78 |
| The Other Half. | 80 |
| The Potential of Stones. | 84 |
| Memories of Gorrondatxe. | 86 |
| Quantum Entanglement. | 90 |
| Entanglement. | 92 |
| Marriage. | 94 |
| Work. | 96 |
| Disordering the World II. | 99 |
| Memories of Covadonga. | 101 |
| Musical Chairs. | 102 |
| And the Stones. | 105 |
| Memories of Stonehenge. | 107 |
| The Void. | 110 |
| Moving Stones I. | 113 |
| The Sea. | 115 |
| Moving Stones II. | 118 |

| | |
|---|---|
| Physics. | 120 |
| Moving Stones III. | 126 |
| Love. | 130 |
| Moving Stones V. | 133 |
| The Stones. | 137 |

# MEMORIES OF SESTAO

If there's something that feels strange when relating to my own past, it's the fact that I'm now aware that I'm doing it from the perspective of being a high school teacher.

*"My physics teacher once told me very seriously, 'Why don't you take up opera? You have a deep voice, like an opera singer.'* So, I decided to study physics.

My relationship with physics is quite toxic. We love each other a lot, we understand each other very well, we need each other, but we also hurt each other a lot. Well, essentially, it hurts me—I can't really do much to it because it's an abstract entity—or not even that, it's just the name of a science. So, after four years of college and a year of a master's program, I told it that I needed time and space. It understood, of course, although it didn't understand me talking about time and space as independent things. Now I aspire to tell future teenagers that they should take up opera."

ΔΔΔ

"I still remember when, in my first year of high school—or maybe the second, I don't know, I'm terrible with dates—my Basque language teacher told me that my essay on music was too childish. According to him, it lacked reasonable arguments for someone my age. He said that the ideas in the essay were childish, empty, lacking maturity and value. I think that's where it all started. That essay was, for me, a damn declaration of love to what I loved most at that time: music. And I loved music because only it filled that existential void I already had deeply rooted.

The truth is, I don't remember if the essay was childish or not, or if it was really written with empty words. The latter is very possible, as I've always been terrible at writing about what I love the most. The greater my feelings, the worse the rhetoric that has always accompanied them. Probably because, feeling that any word would be insufficient, I prefer to abandon any pretension and let what I'm describing—or who I'm describing—shine through their own existence. And it seems that, for many people, that's not enough, the simple fact of existing. We value existence, pure existence, mere existence, very little. If at all. And, I don't know, for me, there is nothing of greater value than the simple fact of existing, both oneself and everything that surrounds them. After all, it's the only stone I've moved all this time, so how could I not value it?

Going back to my essay on music, I remember writing about how it helped me escape from reality and problems. And how ashamed I am to have written that. I would never say such a thing now. First, because it's not true, and second, because it's not something to be proud of. We abuse escapism a lot, don't you think? We drink to escape, watch TV to escape, consume junk online to escape, have sex to escape, sleep to escape, play video games to escape... and yes, we listen to music to escape, too. What a horror life must be to require constant escape, right? Thankfully, that's a lie. Maybe that's what my teacher meant by calling it childish—believing that music could help you escape from reality or, even worse, wanting to escape from it, goodness, what nonsense. If I had to write that same essay now, I would put the complete opposite in it, for sure. I would write about how much music has taught me about reality."

∆∆∆

If there's something undeniable, it's that the story of my past, like the story of so many others, I'm afraid, is the story of someone who, faced with a feeling of not belonging to the established order, has had to constantly dedicate all the power of their mind —and their being—to generate alternative narratives to hold onto. Narratives filled with insecurities, always. Often, unconfessable narratives. Narratives to keep surviving. Narratives to be able to relate to, against, and in the face of the hegemonic narrative that subjugates us.

These narratives, with more or less power, are all equally necessary for us. And they often arise from the most seemingly unexpected, frivolous, or superficial places. Like that essay on music I wrote. I don't know what I put in it, and it's very likely that the teacher—whom I don't blame for anything, of course— was entirely right in their criticism. But at that moment, what I do know is that that essay was one of those narratives for me.

Years later, playing some video game or other, one of the commentators on the in-game radio said the following phrase:

"Pop music doesn't judge you."

That simple phrase resonated in my head, bringing that damned essay on music back to my mind. That was it. So simple, so clear. That's all I felt with music. That was all I needed to express about music at that moment. And about music. And about anything else that exists.

JAITZULF.

◊~◊

. This book is nothing more than an amalgamation of thoughts and ideas that my mind has transformed into an eternal present. This book is merely a collection of chaotic and disordered ideas upon which I attempt to impose an apparent order.

◊~◊

# THE VOID.

JAITZULF.

The idea of nothingness, the absence of everything, absolute silence. Or, well, maybe not. What if I told you that the void isn't really empty? What if I told you that, where there is nothing left, what you find is absolute chaos?

Quantum physics, the theory that describes the foundations of the world, tells us that the void is actually a seething hub of activity. According to this theory, the void is a sea of particles and antiparticles that appear and disappear in the blink of an eye. They come and go so quickly that we can afford to pretend nothing is happening—just quantum fluctuations. The void is a place of infinite possibilities. It is, in fact, where the reality we know takes shape from nothingness. Just as a blank canvas is the beginning of a work of art, it is believed that the quantum void is the origin of everything that exists. Every particle, every atom, every star, every galaxy, every kiss, every song, every drop of water, every hug, every stone... all of it emerged from this sea of possibilities. And just as a canvas is never truly empty, the quantum void is also always full of potential—of incredible potential for beauty and creation. So we must ask ourselves, what does this mean for us?

The greatest revelation I had during my years studying physics at university was precisely this idea that the void isn't really empty. Discovering that the void is actually just the appearance of nothingness, hiding behind it an absolute chaos of interactions so rapid, fleeting, and self-destructive, was both the onset of a crisis in the absolute beliefs that had formed the foundation of my thinking (how could the void not be empty?) and the beginning of building a new framework of thought to reconcile my mind with the world. Tearing down the walls we construct in our minds to feel comfortable and secure is undoubtedly terrifying, but it's the only way we can truly enjoy the world beyond them. And the truth is, the world that opened up to my mind once I let that wall fall— the wall that so strongly protected the classical idea of the void—

has never ceased to amaze me. It turns out life is better outside the wall, even if it didn't seem so at first.

But why did I find it so terrifying to let go of the idea of nothingness? Honestly, I'm not quite sure anymore. After all, the definition of the void that physics gives us is exactly the one I've always felt in the depths of my being: an absolute chaos of interactions so rapid, fleeting, and self-destructive that they give the appearance of nothing happening. That's how I've always felt the void: like utter chaos—utter chaos that pretends to be orderly, to be exact.

Well, let's not cheat; we were talking about the concept of the void in physical space. In that case, and thinking about it now with perspective, it doesn't really make much sense for letting go of that idea to be such a big deal. Has anyone ever experienced physical emptiness? Has anyone been able to observe it? What we understand as emptying something in everyday life is actually replacing the contents of something with something else—usually air. For example, emptying a glass of water is simply filling it with air. And while outer space is often considered a vacuum, the reality is that the entire universe is full of stars, gases, dust, cosmic radiation... The appearance of emptiness and the need to generate a contrast to everything we know is what has given rise to our idea and concept of the void, not the empirical observation and experience of actual emptiness. In fact, are we even capable of imagining emptiness in our minds? If you try, you're most likely going to picture an empty box (or something similar)—a box that would fill with air and be penetrated by cosmic radiation as soon as you placed it in the real world, which, in itself, demonstrates our inability to imagine nothingness: you're imagining a box, which is already something, and therefore no longer void.
The reality is that we've never experienced the void. Not even our minds are capable of imagining such a thing. Nothingness never existed. There was never a void. At least not in our experience.

JAITZULF.

And yet, we base everything on it. Our entire construction of the world, of our knowledge and collective imagination, stems from filling nothingness, from filling our voids. Well, perhaps it's time to start building from somewhere else. How about we try building from the sea, from physics, from love, and from stones?

# AND THE SEA.

JAITZULF.

*The past is always new,
the future is always nostalgic.*
Read a neon sign at the Museum of Modern Art in Rome.

# MEMORIES OF MURCIA.

One of the memories I cherish most and recall most often is from a school trip to Murcia when I was in fifth grade. They took us to the beach to gaze at the stars, as far away as possible from light pollution. That memory from the past is so recurrent in my mind that I could almost say it has stayed with me in an eternal present. What's curious about that memory is that when I looked up at the sky to see the stars and dream of the constellations that the guide was describing, everything we were seeing and experiencing as a fascinating present was already, at that moment, coming from the past. A very, very distant past, in fact. You've probably come across this idea before, but it's always worth remembering that the stars are so far away that by the time their light reaches us, by the time we're able to see them, many, many years have passed, and some of them may no longer even exist. Looking at the sky on a starry night is fascinating, especially if you do it in a place far from the city lights. Looking at the sky on a starry night is looking into the past, no matter where you do it from. In fact, it's looking into different pasts. If those stars were observing us—unless they were among the unfortunate ones that have already collapsed, of course—some might still be seeing dinosaurs on Earth, while others might be witnessing the Earth forming or even not yet starting to exist. Everything I perceived that night came from the past—not just the light from those distant stars, actually; everything else I was seeing, too. Just like what I'm seeing now:

everything we see is always images from the past. Although not always from such a distant past, of course. Normally, they come from a past so infinitesimally small that we can take it for the present.

We can apply this same idea to all the other sensations I was feeling at that moment on that beach by the Mar Menor: the instructions given by the guide, the shouts of my classmates, the noises made by people walking along the beach... all those sounds took some time to reach my ears from the moment they were made. In fact, each one took a different amount of time to get there (depending on how far away they were and what was in between).

In reality, that magical ensemble of constellations in the sky, scented with the deep smell of the sea and accompanied by a variety of sounds, that scene that I remember as a fascinating present which my mind recorded and has kept alive to this day, is nothing more than a collage of different sensations originally generated at different times. The memory of that present is a collage of different pasts that my brain converted into simultaneous events by processing them all at once—and that has since continuously altered them. That scene from the past is as present now as it was then: it only existed in my mind, and it will continue to do so as long as it remains stored in my memory.

In the same way, by the time this book reaches your hands, a long time will have passed since these thoughts were thought, processed, and translated into words. Each of them will also come from different moments of writing. Everything in this book is already past—it always was. A collage of different pasts, like everything we perceive. It's likely that I no longer feel, think, or even recognize many of the things written here. At least, that's what I hope.

# THUNDER AND LIGHTNING.

The idea that looking at the stars is looking into the past—a collage of different pasts—is, like so many other ideas science reveals to us, something we usually relegate to the drawer of "magical phenomena of the universe" and pretend that it belongs to a reality outside of us. It's fine to think about on a night gazing at the sky with the love of your life—or of your present—but not when heading to work. Thinking that everything you perceive takes some time from when it happens as a phenomenon until you perceive it, and that the things you perceive at any given moment don't necessarily come from the same prior instant, makes no sense when in daily life 99.99% of what you perceive does reach you practically instantaneously. However, this idea that the things we perceive are often out of sync is nothing new, nor entirely foreign to our everyday life. And it's not a magical phenomenon exclusive to the stars. In fact, there's a physical phenomenon we are familiar with that makes this very obvious: lightning. Everyone knows that there's often a noticeable delay between when we see lightning and when we hear the thunder that follows. This allows us to estimate how far away the lightning struck by doing something as simple as counting the seconds between the lightning flash and the sound of thunder, then dividing by three (which gives us the approximate distance in kilometers). It also reminds us that bright stars in the sky aren't the only things we perceive out of sync. We see and hear the same phenomenon at different moments, just as we see stars from vastly different eras when we look up at the clear night sky. And this applies to everything, at all times. Everything we perceive takes a certain amount of time to reach us. And not everything takes the same amount of time. Nor does everything that happens even reach us, of course.

This leads to an inevitable conclusion: the perception of time, or at least the perception of the present, is an experience that is inevitably tied to the observer. It's tied to each person—and beings and things in the world in general—that experiences it. It's an individual experience. Each of us—not as conscious beings, but as entities that exist—perceives the present and constructs it from similar information, if we're relatively close and in similar conditions, but never the same. Obviously, our cognitive abilities allow us to reconstruct events and arrange them in their proper order. "My senses deceive me, but I know that lightning and thunder are two sides of the same phenomenon that happened simultaneously." Alright, alright, no need to get upset. Of course, they are. But in any case, the supposed ability to abstract ourselves from our perception and organize events jointly has nothing to do with what I'm trying to show here, which is that the present, if it exists at all, is merely a collection of perceptions from different moments in the past, united in a single temporal moment only by our mind. Whether you're looking at the stars, watching lightning, watching TV, or catching a knowing glance from your crush at the back of the club. Nothing reaches you instantly, nothing arrives at the same time everywhere, nothing is present for the entire universe simultaneously. In other words, nothing is simultaneous. There is no present shared by all. And yes, I know that in many cases these differences are imperceptible to us, but that's exactly where the true trick of your senses lies. Your senses don't deceive you by showing you that you see before you hear, they deceive you all the rest of the time, when your mind's processing erases that imperceptible delay in the information that reaches you, making you believe things are happening simultaneously. As if the entire universe moved forward in sync. As if there were something, I don't know, called time, external to us and universal to all, forcing every single one of us to move forward together.

# SHARING THE PRESENT.

*"I don't know if you've ever heard that when you touch another person, there isn't really any physical contact as you might have imagined all this time in your head. There aren't things physically colliding with each other. What really happens is that your atoms and the other person's atoms are interacting electromagnetically; they don't actually touch—in fact, they repel each other. If we go a little further, what's happening is that your body and the other person's body are exchanging virtual particles through the void—virtual particles, mind-blowing stuff. And you don't just exchange them with someone when you touch them, but you're exchanging them all the time with everything around you. Physics teaches us, then, that nothing really touches anything, but that everything, absolutely everything, is in constant exchange, in constant flow. I don't know about you, but for me, this makes touching someone even more magical. Try touching someone—their hair, their hands, their ears—their damn ears—their cheeks... while imagining that the tiny particles making up your fingertips are exchanging little packets of energy with the particles of that person's skin in that very moment. Try it—with someone who gets you, ideally—and tell me it's not one of the most wonderful sensations in the world."*

I wrote that when I was 25 years old, under the influence of alcohol. It's been more than five years since then, and honestly, touching people still fascinates me just as much, if not more. When I wrote those lines, I won't lie, I wasn't the kind of person who gave hugs or showed affection through physical contact. In fact, I found it annoying. And I still do, depending on who's doing it, of course. Reflecting on these words again, and after all that I've lived and learned in the past few years, I've recently come to a new conclusion: I think the sense of well-being that comes from caresses, hugs, kisses, and all other expressions of love has a lot to do with the simultaneity we were talking about. If our life experience is our mind's interpretation of signals coming from different pasts—gathered into a present that is unique to each of us—and if the perception of time is inevitably tied to the observer, then simply touching each other, hugging each other... these are experiences where we can be certain that we are sharing simultaneity with another being. Because they are the only directly shared experiences. Physical contact is one of the few truly simultaneous experiences, one of the few shared presents, that we can be sure of.

In the end, maybe all the narratives created around the magic of love are true, and the pleasantness of the experience simply comes down to feeling unified, to feeling in the same present. Hmm... well, no; the idea of exchanging virtual particles is still mind-blowing, just thinking about it.

# DISORDERING TIME.

Time, huh. What a fascinating thing. What is time, really? I don't know. Surely, no one knows. If there's one thing Einstein made clear, it's that time is relative, right? Well, yes and no. Like everything. Everything is yes and no, I guess. That's what relativity is about, isn't it? That everything is relative. Well, not really, quite the opposite, but we'll get to that.

The definition and perception of time, surprise, depends on who you ask. It's clearly different for people working in different fields. Time isn't the same for a psychologist as it is for a neuroscientist or a physicist, for example. But it's not even the same for two physicists studying different areas or theories of physics: Newton's time, Einstein's time, and Schrödinger's time are different things. And don't even get me started on how long it's taking science to truly recognize women. But I suppose that, like everything, that's already in the past—or maybe not? Anyway, one thing we can all probably agree on is that any of these definitions or conceptions of time help us organize events in our minds so that we can know what happened "before," "now," and "after." Or maybe not?

Certainly, in Newton's case, this is an absolute truth. Like everything in classical mechanics: an absolute truth. Classical mechanics views every variable, including time, as a fixed value. There's no room for improvisation. Once the characteristics of a system are known at a certain moment, classical mechanics tells us that absolutely everything—both the before, the now, and the after—is determined by the physical laws that govern the world. This is quite a logical way to understand the universe, as it apparently aligns with our daily experience of reality. All the events we live through—and share with other people, beings, and objects—seem to follow a clear temporal sequence, and with some knowledge of physics and math, we could explain them without much complication. Once the initial conditions of a system are known, classical physics gives me an equation that tells me how it will evolve without much trouble.

But.

Einstein completely shattered this image when he studied what happens to objects moving at very high speeds (close to the speed of light, the limit of our reality). And, as incomprehensible as it may seem, for objects moving very fast, "now" becomes a fuzzy concept, harder to define. The relativity of simultaneity is one of those concepts in modern physics that's hard to grasp because it completely contradicts the worldview we've created and everything we experience daily. However, the reality is that two things that seem to happen simultaneously to you don't necessarily happen simultaneously to another observer moving differently from you. In other words, your "now" and that other observer's "now" can be different if either of you is moving at speeds close to the speed of light. What's more, an observer at rest and one moving at high speeds don't have to agree on the order in which certain events occurred. And it's not a matter of perception; they truly didn't happen in the same order for both of them. Wow. On top of that.

Quantum mechanics, the most successful physical theory to date, discards time as a variable and demotes it to a parameter. In other words: it doesn't give it the status of an "observable." And what isn't "observable" doesn't play much of a role in the quantum world, because that basically means it can't be measured. In quantum mechanics, we can talk about whether a particle exists or doesn't exist in a certain place, but not whether it does so at a specific moment or another. In a way, time in quantum mechanics is just a tool that pushes the system it describes to progress. Whatever "progress" means.

So.

The socially accepted and understood concept of time has long been dismantled by science. Over the last century, both quantum mechanics and general relativity have debunked the ideas about time that classical mechanics brought to the world. Ironically, despite being one of the culprits for cementing that idea socially, even Newton himself didn't believe that the concept of absolute, universal time, external to the system, was real; he saw it simply as a tool that worked. Just like quantum mechanics does now. However, nothing is more uncomfortable for any human being than admitting that the basic principles they've built their thinking on might be wrong. Even more so when these principles are actually deeply held beliefs. How can the simultaneity of two events not be a concept shared by any pair of observers if it's what I've been experiencing my whole life? How can time not be something measurable and observable if my entire life has been ordered and structured since I can remember, based on what a clock says? Obviously, it's a hard pill to swallow because it defies all logic and previous experience. Luckily, the fact that the simultaneity of two events is not a shared experience only has notable effects if there's an observer moving at speeds close to the speed of light, and since a human reaching those speeds is highly unlikely, we can keep pretending that our concept of simultaneity is correct for us. Instead of redefining our vision or accepting that it's limited, we can just say that those effects only apply in certain conditions. In other words, we can relegate those effects to something out of the ordinary and carry on as usual. The same goes for the idea that time isn't an observable in quantum physics. Or for queer people.

Quantum physics is probably the theory that suffers most from this: everything that scares us about quantum physics, we relegate to simply being a "quantum effect." Whatever a quantum effect is. If every idea that doesn't fit—apparently—with daily human experience is usually stashed away in the drawer of "magical phenomena of the universe," I would dare to say that the entire theory of quantum mechanics has practically been stashed away there since day one. Everything that scares us about the theories that came to kill God is relegated to resources for fantasizing, for fueling science fiction or the hopes of chaotic minds: let's speculate about futuristic spaceships, time travel through wormholes, and millions of parallel worlds, but for God's sake, don't let anyone accept the idea that there's no privileged observer and that we're just puppets in a space-time framework! Never!

# TEARING APART THE SEA.

To observe the sea with our eyes is to glimpse an enormous and continuous mass of fluid, incapable of remaining still. The sea is a continuous flow before our eyes. Like time. The sea behaves as if it were a single entity, a single reality, moving together in a constant struggle to conquer everything that is not the sea. Like time. The sea is a continuous flow trying to break down walls, cliffs, moving and destroying rocks, forming grains of sand. Like time.

The sea is the sum of many things. The sea contains an immensity of things within itself. An infinity, one could say. The sea contains so much that even today, when we consider ourselves to be perfect knower of our planet and dare to define even that which lies beyond its limits, we still know very little about what inhabits the sea. We live, in large part, detached from the sea. As if we even feared discovering its secrets. And it's not surprising, because who dares to tear apart and organize the sea?

No one looks at the sea thinking of it as trillions of drops of water, even though it is. No one dissects the sea with their gaze. It would be an immense and absurd task. None of those drops has identity or relevance without all the others surrounding it. None of those drops is the sea on its own; it's just a drop. However, if we try to understand it in its entirety, if we try to understand every one of its characteristics, we cannot ignore the fact that that's exactly what it is.

Trying to create a mental image of the sea as the sum of its basic components is a titanic task. At first, one might think that the most abundant and defining component would be the water drop. The smallest possible unit of reality for understanding the sea would seem to be the water drop. And, well, tracking water drops can help us understand the movement of tides, waves, the formation of tsunamis, and other marine phenomena. But that would be insufficient. We would soon realize that the water drops are made up of a number of water molecules that I won't even bother to write because it's inconceivable to comprehend. And those molecules, we know, are made up of hydrogen and oxygen atoms. Atoms that, we also know, are made up of electrons, protons, and neutrons, and the latter are, in turn, made up of quarks. And that's just talking about water, let alone all the other things in the sea!

What's fascinating is how, somehow, the sum of all those things ends up appearing as a continuous whole. What at a fundamental level is a bunch of jumbled particles interacting chaotically with each other, from our worldview, becomes a single perfectly unified, cohesive, and continuous entity that also wields incredible power. Like time.

# MEMORIES OF ORIÑÓN.

One of the most beautiful phenomena that nature gives us is the tides of the sea. Near where I grew up, there's a beach—in the small coastal town of Oriñón—that used to seem immense to me. When I was young, my parents, my grandfather, and I would regularly go there on summer weekends to spend the day. Chairs, beach table in the trunk, cooler filled with ice, and containers of food my mother had cooked in the morning. Food always tastes better outside the house, but eating in a natural setting always tastes even better, even when the wind lifts the sand and makes the paper napkins fly.

What fascinated me most about that beach was that I never knew what I was going to find. It could be an enormous body of water with a small strip of sand where you had to squeeze in with many other families, or it could be the opposite: a huge expanse of sand cut through by a river with the sea far in the distance. Two such different and equally beautiful landscapes belonged to the same place. And the Moon decided which one to show. At a certain age, I began to understand that those tides happened regularly, influenced by the Moon, and that, by simply looking at the newspaper, I could anticipate what I would find.

For many centuries, it was the perception of changes like these that shaped the idea of time. The day-night cycle, the seasons (defined differently in various times and places due to differences in perceived changes and cycles), the tides... The farmer would rise to work the land when the rooster crowed because that meant the rooster had perceived the light of the sun and thus knew that there was light to work by, not because it was five, six, or seven in the morning. He didn't even have that concept. "The sun has risen again," the rooster would say. "I have light to work," the farmer understood. It wasn't until the advent of industrialization and its imposition of a productive mentality that time was established in more abstract fixed units (hours, seconds) that allowed for the simultaneous organization of human life. Until that moment, time in certain cultures was even understood as cyclical (day-night, winter-summer, high tide-low tide) rather than linear. The need for linearity first arises from the need for narration—historical, religious—that becomes much harder to comprehend without a beginning and an end, and later, from the aforementioned productivity: everything must flow forward and never stop.

With the changing perception of time came changes in how it is measured. Technological advances and societal organization have led us to completely abstract both the concept of time and how we perceive it, which has, in turn, shaped society, closing a cycle as time used to do. The perception of the cycles of the stars (the sun, the moon, the rotation of the Earth itself) gave way to programmed machines that replicate these cycles at their own pace: clocks. And with them came the ability to divide time into hours, minutes, seconds, and any other units we wished. Although the idea and definition of a day originally came from perceiving the Earth's rotation and was taken as the complete time it took for it to make a full rotation on its axis, today, time is no longer scientifically defined by that rotation. The reason is that the Earth doesn't rotate at the same speed all the time; in fact, it rotates ever so slightly (very slightly) slower, so it's not a definition we can use for the precision we now require. For this reason, the unit that is scientifically defined today as the reference and foundation for the idea of time is the second. And the definition of a second is now "the duration of 9,192,631,770 oscillations of the radiation emitted by a cesium-133 atom when it transitions between two specific energy states." Wow! That duration is almost the same as if we measured the time it takes for the Earth to complete its rotation, but not exactly, and it will become increasingly different as time goes on (a lot of time for a human, don't worry). In any case, socially, seconds—like minutes and hours—are just numbers ticking forward that make us, on the one hand, more organized and less unpredictable, and on the other, more enslaved to the productive mentality. Tick, tock, tick, tock. Produce, move, do things. Funny how the rooster used to tell the farmer the same thing.

# CH-CH-CH-CHANGES.

As I was saying, humans didn't have a linear perception of time until relatively recently. The notion of time comes from perceiving the changes that happen around us and finding patterns in them. It's not the perception of the random changes we constantly perceive that gives rise to the concept of time, but the repetition and recognition of cycles in those changes that allow the establishment of the idea of day, lunar cycles, tides, seasons, etc. If the changes we perceived didn't follow any kind of pattern or cycle, the concept of time would probably never have emerged. What I'm saying here actually applies to all knowledge, not just time: it's the patterns, repetitions, similarities, comparisons, and reiterations that give rise to understanding. A succession of random data is in no way interpretable by our minds. And, to our misfortune, our reality is in constant change. Everything around us is changing. To a greater or lesser extent and in ways more or less perceptible, but everything, absolutely everything, is changing. However, it is precisely the fact that some things change in seemingly predictable ways and that there are things whose changes are imperceptible to us that allows us to discern changes in everything else. It's in stability and in cycles, in order and the ordered, that our mind finds comfort and dares to define. Always resorting to the same trick of establishing around that cycle and stability the center of normality and the reference from which to define everything else—to define disorder. At least until those things (disorder) become something recurrent to it (order). And that's how days and seasons gave rise to years, hours, and months. Based on detected patterns—lunar and solar cycles, essentially—other concepts were built that don't fit an observed physical reality but allow us to organize and establish order in other concepts born from other sources of knowledge, such as politics or sociology. I mean, just as being aware of solar cycles allows us to organize work in the short term (day-night, light-no-light) and long term (seasons, harvests) to live better, organizing life as a succession of events that allows us to learn from the past and build a better future is also an advantage for the survival of the species. And it becomes inevitable once we begin to craft narratives—political, religious, philosophical—that attempt to understand that past and give meaning to the present as the place to plant the seeds of the future. But just because it's inevitable in the organization, first of our minds and then our lives, doesn't mean it's physically or essentially accurate. It simply means that it's a useful tool for organizing ourselves. An indispensable tool for ordering the world so that we can govern it. But not for knowing the physical reality behind the concept; for that, we may need to try to strip away all the constructs that made it linear and return to what planted its seed: the detection of patterns in the perception of physical changes.

# COSMOS.

The word "cosmos" comes from the Greek -κόσμος- and means order or adornment. The idea of the cosmos is the counterpart to the idea of chaos: the notion that the universe is orderly and harmonious. In ancient Greece, the universe was perceived as an orderly, structured place, far from chaos. Astronomical observations, the harmony in their movements, and the ability to predict those movements gave rise to this idea. The predictable motions of celestial bodies, the phases of the moon, the seasons, among other phenomena, suggested the existence of an underlying order. They suggested the existence of an ordered and harmonious truth from which to understand the world.

For a long time now, humans have categorized the universe as an ordered place. This idea of "order" has survived for many centuries, partly because of the way our minds understand the world—ordering is the first step to understanding—and also because it aligned with dominant beliefs and hegemonic thought influenced by prevailing religions. Purity, perfection, and absolutes dominate our way of understanding the world. Even if we don't want to, even if we don't believe. Perhaps that's why, even when scientific advancements have made it abundantly clear that this worldview is completely false, and that behind this appearance of order lies absolute chaos, we still talk about the "cosmos." In reality, the universe is an immense chaos: the result of an unimaginable number of interactions. Interactions that, from our understanding, are of a probabilistic nature. The universe is a sea of possibilities. A collection of interactions between different entities that occur and are possible simply because they could occur and could be possible. And although there are virtually infinite interactions happening at every moment, probabilistically, and with a multiplicity of possible apparent outcomes... we are beings capable of organizing all that information in our minds to the point that we've decided to name the world "cosmos." Wow. We are truly fascinating beings.

# THE ARROW OF TIME.

If there is one principle in physics that seems to withstand any test thrown at it, it's the second law of thermodynamics. It's a fairly simple principle in appearance: "the amount of entropy in the universe always tends to increase." But appearances can be deceiving, and with this simple statement, we've been able to understand things as varied as the combustion engine that powers cars or black holes. Of course, what the heck is entropy?

Entropy is a measure of how disordered the universe is from our perspective. It's a concept that arises from our inability to distinguish between different equivalent states from our viewpoint, but that can be radically different if we zoom in at a microscopic level. Think, for example, of a pile of sand on the beach. At first glance, all the grains look the same. However, if we could look closely, we'd see that each grain has a unique position and orientation. If we change the position of the grains, the pile would still appear the same to us, but in reality, at the microscopic level, every arrangement is different. Entropy reflects this lack of knowledge. From our perspective, the sand on the beach has more entropy than the rocks from which it originated. The grains of sand can be rearranged in countless ways without changing their macroscopic appearance: they remain a pile of sand. Rocks, on the other hand, are more defined structures; if the parts that make them up were to move, the rock itself would change in appearance, size, or even consistency. In other words: rocks are a more ordered state, while sand is a more disordered one. So, what does it mean that the amount of entropy in the universe always tends to increase? Basically, it means that the universe is constantly tending toward disorder. And something becoming more disordered, as we've seen, implies that our knowledge of it decreases. The conclusion is clear: the universe is not only chaotic, but it's also a chaos that becomes increasingly disordered and, with that, increasingly incomprehensible to our eyes. Wow. So, what now? Well, we'll just have to keep disordering.

The concept of entropy, the amount of disorder in a system, is also the only thing that dictates the direction of time in the universe. It is the arrow of time. No other theory specifies a single direction in which time can move. To say that time moves forward and to say that the world is becoming more disordered are, essentially, synonymous. The passage of time is the disordering of existence. No equation where the variable "time" appears in any physical theory indicates the direction in which it must advance. The only equation that pushes the world in a definite direction is none other than:

$$\Delta S \geq 0$$

That is, the increase in entropy. "Delta S is greater than or equal to zero" is how that equation would be read. Where Delta is the uppercase Greek letter used in science to represent change, and S is entropy. Time moves forward and never backward because the entropy of the universe tends to grow and never decrease. Time moves forward because the world becomes disordered before our eyes.

It's curious, isn't it? As humans, we've built the concept of time by perceiving order: patterns, cycles... and yet, time itself is born from the world's tendency to become disordered before our gaze.

# DISORDERING TIME TO ORDER THE WORLD.

One of my biggest obsessions during my time studying physics was the idea that everything, absolutely everything that exists, is a broken symmetry. You only have to look in the mirror to understand this idea: if you divide your body in half from top to bottom, what's on one side and the other is practically the same, but it's not exactly identical. The difference is subtle but noticeable enough that you can easily distinguish between your image and your mirrored image, even to the point where it could give you nightmares. Everything in the world is like this, essentially. Nature itself has an incredible tendency toward symmetry, but it always presents it in a broken way. Everything is symmetrical-but-not. From animals, trees, leaves, minerals, rainbows, the structure of DNA... or the structure of the universe itself. And it's a good thing, too, because otherwise nothing would exist.

Among the many broken symmetries that give rise to the world, time is clearly one of the most prominent. Whatever time is and whatever its interpretation, no one, ever, would interpret it as being able to move in the opposite direction from the one it does. No one has ever experienced or perceived time flowing backward. Yet no physical law explicitly states this. Newton's, Maxwell's,

Schrödinger's, Einstein's laws... all are symmetric with respect to time (and completely break symmetry with respect to gender, by the way). In all of them, it doesn't matter if time moves forward or backward; they work the same. Although the world doesn't. Clearly, whatever the symbol "t" represents in those equations, it's something very useful, but not fundamentally accurate. It's clear that the need to organize the world before our eyes using absolute, perfect, and incredibly powerful mathematical equations has led us to assume as known, fundamental, and understandable a concept that we've blurred and stripped of all real essence.

We've disordered time to order the world.

# AND PHYSICS.

JAITZULF.

*"A man can be destroyed but not defeated,"*
*—said the foolish Hemingway in The Old Man and the Sea.*

# NEWTON WAS UGLY.

If studying physics has taught me anything, it's how to develop, over the course of ten years, a deep animosity toward Sir Isaac Newton.

Newton, known as "that gravity guy, the one with the apple," is commonly presented as one of the fathers of classical physics. Rightfully so—after all, he formulated the famous laws of motion, formally known by his name: Newton's laws. These, along with his famous law of universal gravitation, had an enormous impact on our understanding and perception of the world. For the first time, humanity was not only able to describe and predict the motion of the objects around them, but also to describe and predict the motion of the celestial bodies they observed as bright points in the sky. And all with the same formulas and calculations! With him, celestial objects ceased to be divine objects and became as mundane as those on Earth. Or perhaps, better said: the terrestrial became as divine as the celestial.

And you might ask, do you hate Newton for that? The answer, obviously, is: no. I hate him for something much deeper: his law of inertia gave rise to an entire conception of physics that kept me up at night for years. The law of inertia (also known as Newton's first law) states that:

"An object at rest will remain at rest, and an object in motion will remain in motion at a constant velocity and in a straight line, unless acted upon by an external force."

Starting from this law, whose content I won't dispute because it's evident and more than proven, a whole worldview of physics—and therefore of the world and/or existence itself—was built, in which everything is understood as a sum of isolated entities. As a sum of individuals. As a sum of rocks. This law—along with others—gave birth to the cold world of classical mechanics, where everything that exists is understood as parts of a machine, as individual pieces that together form a grand mechanism. Chilling, if you ask me.

Moreover, his figure has been used to build an entire narrative around intelligence and the discovery of knowledge that I've always found unbearable. The truth is, an apple never fell on his head. The story that one day, while seeing an apple fall, he came up with the law of universal gravitation is nothing but a myth, created to glorify him as a genius. Personally, I believe there is nothing less educational or more counterproductive than telling new generations that great ideas are born from coincidences or innate gifts. As if Newton didn't owe his ideas to hours upon hours of time spent thinking and discussing them. As if his ideas didn't come, like everything else, from his constant interaction with the world. As if Newton were really just a cog in the machine destined to keep it running because, in some way, he was designed for it.

Oh, and then there's his view of absolute space. Seeing the world as if it were a giant machine requires that this machine, or at least the parts that compose it, be contained within something. And that something is what we call space. Space, as I've mentioned before, was established as an absolute and infinite entity in Western collective thought largely because of the popularity and impact of classical physics. In other words, thanks to Newton's success. Judging him for it now is absurd, because anyone who had observed what he had and could test the viability of his laws on both Earth and in the sky would have arrived at the same conclusion: space and time must be absolute and infinite. Only then would it make sense for his mathematics to describe the seasons, the tides, and a ball rolling downhill with equal accuracy. However, establishing "absolute" ideas backed by such overwhelming and uncontestable success—like seeing humanity reach the moon thanks to that classical physics, which now seems like I want to disparage—leaves a legacy that's hard to rectify. And maybe that's why Newton has always seemed so ugly to me. After all, Newton taught us that seeing the world as a sum of independent entities exerting forces on each other to transform reality is an incredibly powerful vision. So powerful that, since then, all we've done is continue using our force to tear the world apart.

# TEARING THE WORLD APART.

To order the world, to tear it apart, to divide it into well-defined entities with specific characteristics, and to label those entities... has allowed us to have a consciousness about what the world is and a mastery over it that far exceeds what any other living being could ever have dreamed of. There's no doubt about that. It also brings us the comfort of believing we know how everything is going to behave at any given moment. But it's not the truth of the world. The world is as much ocean as the ocean is world. Imagine trying to understand the ocean by separating it into the 10 trillion drops that make it up and observing every single interaction between the drops to explain the movement of the tides. It would be an immense and absurd task. None of those drops has identity or relevance without all the others surrounding it. None of those drops is the ocean by itself. No one looks at the ocean thinking of it as the sum of trillions of water droplets. Even though it is.

After Newton's success—and the development of classical mechanics in general—science, particularly physics, developed a noticeable obsession with compartmentalizing, specializing, and dissecting its objects of study more and more. At the same time, the obsession with understanding the fundamental building blocks of existence led to the development of the most successful physical theory known to date: quantum physics. Quantum physics—and specifically the branch of particle physics, supported by quantum field theory—is humanity's greatest effort to reach the most fundamental components of reality. The greatest effort to identify the fundamental entities that can be defined individually. But it's not the only one.

The history of physics is largely the history of the search for the fundamental elements that govern the universe. Throughout human history, different cultures and eras have proposed various fundamental elements to explain the world. The Greeks used Earth, water, air, and fire. Some cultures later added a fifth element: the void—or aether. Earth, representing solidity, stability, permanence, and structure. Water as flow, change, and transformation. Air for the immaterial, the invisible, the spirit. And fire as energy, force, passion. Aether, meanwhile, came to fill the voids—both literally and spiritually. It was the element associated with the eternal, the divine, but also the element that was supposed to occupy the empty spaces in the universe. With the advent of modern science, these ideas were replaced first by chemical elements, then by atoms, and finally by elementary particles. After centuries of history trying to break the world down into its smallest parts, it seems that science has finally reached the most fundamental elements: the 17 elementary particles described by the Standard Model of particle physics. Six quarks (up, down, charm, strange, top, bottom), six leptons (electron, muon, tau, electron neutrino, muon neutrino, tau neutrino), four gauge bosons (photon, W, Z, and gluon), and the Higgs boson. Oh, and their corresponding antiparticles if we want to get technical. These are the seventeen particles within which no structure has been found, so they are considered fundamental since they are not composed of anything smaller. Out of these 17 particles, in reality, the vast majority of the matter around us is made up of just electrons and up and down quarks (which form protons and neutrons and are confined within them). The rest of the particles are highly unstable, with very short lifetimes before

JAITZULF.

they decay.

It turns out that the vast majority of the fundamental elements that make up the universe are highly unstable, and yet here you are, worrying about having a crisis every now and then.

# DISORDERING THE WORLD.

Just as Newton came to order the world and set the path for continuing that task, the fathers of quantum physics came to do the opposite: to disorder it, disorder it to the point of making it incomprehensible.

As a math teacher, one of the things I enjoy most is dismantling some of the firm ideas my students have developed by the time they reach their final year of high school. My favorite moment, unsurprisingly, is explaining to them that the well-worn phrase "the order of factors doesn't affect the product" is not an absolute truth, but simply true for those cases where it holds—like all truths. And even though for the vast majority of mathematical operations a human will perform in their lifetime—or for all of them, depending on the person—it's an acceptable truth, you don't have to dig too deep into mathematics to encounter objects like matrices, for which it stops being true. The product of two matrices isn't the same depending on the order in which it's done. Turns out matrices don't commute*. And that's where quantum physics comes in.

$$XP - PX = i\hbar$$

The fact that matrices don't commute scares high school students at first, but they accept it and live with it just as indifferently as they did with the idea that two times three is the same as three times two. The equation written above—the non-commutation of these two specific matrices (position and momentum of a quantum particle)—is the famous uncertainty principle, and while its implications have been and continue to be the subject of debate in the scientific community, it's generally treated the same way teenagers treat the non-commutation of matrices: after the initial shock, with absolute indifference.

Quantum physics was a revolution and sparked heated debates because of the many ideas it brought that challenged the established framework of thinking in physics. A revolution and debate so intense that it ended with a kind of tacit agreement: it established the majority view that the task of physicists was no longer to interpret results, but to obtain reliable, verifiable results. As long as the numbers add up, it doesn't matter what they mean. A resignation to the task of embarking on understanding the disorder. Because, no matter how much we want or try, it's not easy to face the task of reordering a world that seemed so well-ordered.

# ORDERED ELEMENTS.

As I mentioned earlier, for a long time, Earth, water, air, and fire were considered the four components that made up the universe. To think that the world consists of only those four elements might now seem somewhat ridiculous to most of us. It could be compared to believing that the Earth is flat or that water falls from the sky because some deity decided so. However, it's likely that the average person's way of analyzing their surroundings still aligns more with that view than with the idea that there are 17 fundamental particles divided into fermions and bosons—fermions responsible for shaping matter and bosons responsible for governing interactions between them. Moving from a world governed by four fundamental elements to one governed by four fundamental interactions (electromagnetic, strong nuclear, weak nuclear, and gravitational) is far more complex to grasp socially than it is to discover and theorize about. In fact, like many major advances in modern science, it can't be said that it has caused any significant shift in social thought.

Humans are driven by beliefs. It's always been that way, and it likely always will be. In recent centuries, some belief systems have been dismantled—such as the power and belief in the Christian dogma of the Catholic Church, which has greatly diminished in European societies—but not because beliefs were replaced by certainties and facts, but because they were replaced by new beliefs. The world was no longer created by God; it was created by the Big Bang—whatever that was. Numbers, data, and information now govern the world. They are the new God. Even though we've swapped biblical characters for the names of male scientists, we remain believers. The word of these men, their conclusions and interpretations of the world, are now often accepted as dogmas, and any questioning of them is ridiculed as if it were an attack on deities. But I'm sorry to say it, Newton was extremely ugly and had some more than questionable ideas. Like all human beings. And, like the rest, he also believed in things. Reading the writings and theories of many of these scientists when they talk about their interpretations of experimental results is often truly shocking. Many times, when faced with results that contradict their beliefs and biases about the world, their approach is to find fixes, patches, and tricks to restore "the established order." Take, for example, Einstein's introduction of the cosmological constant in general relativity, with the sole purpose of making the universe static and eternal; or geocentrism, defended for centuries purely out of belief; or the general notion that correct theories or models must be "simple and elegant" like angels; or forcing a "deterministic" interpretation of quantum mechanics.

The development and, especially, the interpretation of science are—like everything—inevitably influenced by the beliefs of the societies in which it develops and by the biases of those who narrate and share it with the rest of humanity. Don't get me wrong; I don't mean to cast doubt on the results or the usefulness of well-established scientific theories. Nor am I questioning the capabilities and enormous contributions these men have made to humanity. We only need to look around to see all the comforts, luxuries, and technological advances that work based on this knowledge and make our lives more pleasant and richer in experiences to realize that it is incredibly useful and functional knowledge.

But.

Physics has long been underutilizing its incredible potential to transform human thought. The very progress of science and the development of societies toward quantifying, measuring, and evaluating everything numerically are leading us toward an immediate and absolute dismissal of everything that is not useful, functional, and productive—and of everything that, though it may be, cannot be quantified. Because that's how we interpret science now, as mere data analysis. If the debate over geocentrism (Earth as the center of the universe) versus heliocentrism (the sun as the center of our planetary system) were happening today, as a model to explain the motion of the planets we observe in our skies, we would likely accept both as valid since, numerically, both models can be made to work (though heliocentrism simplifies the calculations and comprehension significantly, which is why it's the prevalent and universally adopted model). In fact, to the eyes of humans living on Earth, the geocentric model is the one we observe and experience firsthand. So why did science fight so hard against the Church to establish the view that Earth was not the center of the universe? Because back then, it was understood that the development of knowledge and science is a tool for liberation, capable of destroying pre-established dogmas and beliefs and forcing progress toward new frameworks of thought. Ironically, there was such a strong emphasis and fight to establish the idea that Earth is not the center of the universe, that we've now reached the point of ridiculing the entirely valid (though more complex and tedious) idea of describing the movements of the Sun or Moon from Earth's reference point. We've replaced one dogma with another. We are still believers. We still cling to our faith. We're still intent on maintaining order.

# BUTTERFLY EFFECT.

Classical physics, as I was saying, left a deep legacy in Western cultural thought. I also mentioned that this is no surprise: classical physics is the cause of the world as we know it. It is the tool that has allowed us to explore and shape the world at will. But while its foundations and image are based on absolute power, it's far from capable of ordering everything.

One of Newtonian physics 'greatest achievements, as I also mentioned in earlier chapters, was its ability to describe both the heavens and the Earth equally well. The law of universal gravitation is that great tool capable of describing as well the fall of an object from a tower as the orbit of the Earth around the sun. And the Moon's orbit around Earth. And the orbits of the other celestial bodies spinning and spinning around their star. True. But there's a catch.

The law of universal gravitation can accurately describe the interaction between two massive bodies, but no more than that. The moment you add a third body, its ability to provide an analytical solution disappears, and approximations become necessary. Why? Because gravitational interaction occurs reciprocally between all present masses. This means that at the same time one body is exerting force on another, it is also experiencing the force of another, which, in turn, modifies the force it was exerting on the first one… resulting in what's known as the N-body problem—the interaction between more than two bodies, regardless of how many N there are—becoming chaotic:

a tiny change in the initial conditions of the celestial bodies can lead to completely different system evolutions. In other words, the law of universal gravitation would allow me to describe the orbits of a solar system where only the sun and our planet exist with precision, but if we take into account all the planets, comets, satellites, etc., orbiting the sun, the calculations become as complex as they are approximate. So how is it that this law is considered a success? Because if we pretend that, indeed, it's just us and the sun in the entire universe, the results given by the law are good enough. Or just us and the Moon. Or Mars and the sun. In other words, the law cannot describe the entirety of reality, the whole of the masses, but it can relate them in pairs. And the sum of those relations, understood as isolated and part of absolute space and time, resembles the reality we observe. They resemble it so closely, so closely, that thanks to its predictions, humanity has been able to land on the moon several times.

The N-body problem is, in reality, just one of many cases where classical mechanics faces a chaotic situation like this and needs to resort to approximations or corrections based on a simplification of reality in order to describe it. In the classical view of physics, which understands these laws as universal and absolute, the world is often seen as a perfect machine that obeys clear mathematical rules, and once those rules and the initial state of the machine are known, it is perfectly possible to predict every single subsequent step it will take. Under this cold view and understanding of the world, as nothing more than the inevitable result of a process strictly governed by these laws, these kinds of chaotic solutions are usually understood under the idea of the "butterfly effect," which essentially says that this machine that is the world is so absolutely perfect and precise that any slight change in its initial state can have unpredictable consequences. Classical physics, despite presenting itself as a universal tool capable of handling the gears mounted on the absolute space and time framework with precision, must first rely on the absolute definition of the universe's initial conditions to set it in motion; otherwise, the universe is absolute chaos beyond its comprehension.

Drunk on success and fascinated by the undeniable triumph of the tools provided by this worldview, humans adopted its deepest postulates as undeniable truths of the universe. Even though they did nothing more than reduce us to a set of disconnected relationships at the mercy of the absolute rule of space, time, and natural forces. In exchange, we are forced to define absolutely everything around us, at every moment, in order to be able to understand, predict, and control it.

# SCHRÖDINGER'S CAT.

If studying physics has taught me anything, it's to develop a deep animosity toward Schrödinger's cat. In this case, I wouldn't dare say the cat is ugly, like I did with Newton, since no cat in the world is ugly, but I do think it's a thought experiment that, at this point, has lost its meaning. First, because it's constantly misinterpreted and has practically become a meme, and second, because I don't believe it really conveys the concept of quantum superposition. At least not from my interpretation of it. If I had to explain that same concept, I would write something like this:

"As a queer person, there's a phenomenon in quantum physics —one of those that's often presented as counterintuitive and unnatural—that I find quite easy to understand. I'm talking about quantum superposition. An electron lives in quantum superposition in the same way I've lived my identity and sexuality since I became aware of them: undefined, open to the possibilities the world offers, and defined solely and exclusively when interacting with—or being observed by—a third party. The electron doesn't need to be anything or define itself in any way on its own; it only manifests and defines itself for others. And just as, for a queer person, that lack of definition only poses a problem in relation to the environment, the same happens with the electron: quantum superposition is only a conflict because the environment (we) need(s) it 'defined.' It's only a conflict caused by the observer's need to impose order."

Humbly, I think that would be a perfect analogy. Though I fear that not many would understand or share these words, so I would continue with something like this:

"Humanity has been fascinated by the double-slit experiment for a century when, in reality, there's nothing all that fascinating about it. It turns out that when you fire an electron at a plate with two slits and no one is observing, the electron behaves like a wave and somehow passes through both slits simultaneously and interferes with itself. But when you observe it to understand this phenomenon, the electron starts behaving like a particle and passes either through one slit or the other. I think it's pretty obvious by now where I'm going with this, but what happens here is exactly the same thing that happens to queer people in society. The canonical explanation for the electron's behavior and its ability to interfere with itself is that, through some magical and not fully understood process called quantum superposition, electrons can exist in more than one state—position—at once, which allows them to interfere with themselves as long as no one is observing. When observed, that superposition collapses into a single defined state, and that's why we see it behaving like a normal particle, passing through one slit or the other, and it no longer shows interference. In both cases, the conflict doesn't arise from the electron being confused or queer people being confused, but from the observer's imperial need to have a definition. The need to define a person in terms of gender and sexual identity in order to understand them comes from first having created that mental framework where those categories are seen as defining—and not because they are essential or fundamentally defining categories of people. Queer people are only socially conflictive—and experience that conflict personally—because they clash with an order pre-established by a third party, because they interact with conscious beings who store information and impose a certain order by which they understand the world, not because they themselves are inherently so. In the same way, the fact that the electron has undefined characteristics when it's not interacting with anything is only a problem because, in this case, the observer (us) is a conscious being with a pre-established order (either you're a wave or you're a particle) from which they need to categorize the electron to comprehend it. In truth, I lied earlier and exaggerated for effect, but this doesn't mean the experiment and its results aren't fascinating: the issue is that they aren't fascinating because of the electron's behavior as it's usually presented, but because of the implications this has for the importance of the observer and their relationship with what they observe in shaping their understanding of the world."

Finally, as a conclusion to this chapter, I would say:

"Just as the responsibility of not conforming to the established order should never have been placed on queer people—creating conflict and an entire narrative about closets and coming out—neither should the electron have been understood as a conflicting element, nor should narratives have been created about boxes and cats that are simultaneously alive or dead inside them. There has never actually been a cat that is both alive and dead at the same time, waiting inside a box for someone to open it and define it, just as no one has ever really lived inside a closet—only the consequences of others 'need and imposition to define them."

When Schrödinger's cat was introduced as a thought experiment, it was meant to ridicule the prevailing interpretation of the quantum superposition phenomenon—the ability of subatomic particles to exist in more than one state simultaneously when unobserved—by applying it to the macroscopic world, to things our size. However, paradoxically, over time it has become a way to illustrate (without ridicule) the phenomenon. I fear that, in reality, Schrödinger's cat says more about our need as observers to assume that everything must be defined according to the parameters we understand than it does about the cat's "quantum" state. In this thought experiment, it is the cat that truly defines us as observers, not the other way around.

# THE OBSERVER.

Leaving the cat aside, when we say that it's "the observer" who defines a system in quantum phenomena, we must bear in mind that we're not referring to you or me as conscious beings who observe; it's not human consciousness that defines the quantum system. Reality doesn't require a developed intelligence or consciousness to be defined in those terms. It's the interaction with any other element of reality that defines any phenomenon, that collapses the state of superposition, even if it's only defined for me insofar as I observe it directly or indirectly. In other words, when we talk about "the observer," we're simply referring to any element of reality interacting with another. Assigning one the role of "observer" and the other the role of "observed" is, in fact, a narrative constructed from our perspective, because in our eyes, one element is already defined—the measuring instrument—and the other—the electron or particle under study—is not. However, if we were to delve into the most fundamental level and analyze the interactions between the particles involved, we wouldn't find that one has a privileged position over the other.

Translated into the macroscopic world, this concept—that interaction gives rise to definition—would mean that a table is not first a table and then perceived by the world, as classical mechanics would suggest, but rather the interaction of the table's components with each other—thus defining each other as part of that table—and with the rest of the environment—remember that in reality, there are no isolated things; we are always affected by our surroundings—is what defines the table as such. The table is essentially defined by the relationship between the particles that compose it with one another and with the rest of the surrounding universe.

If we return to the world of the smallest things, the indeterminacy and uncertainty we encounter is actually something to be expected. The particles that make up the table, due to their interactions with each other, have established—prior to my interaction with them—an order and relationships identifiable to my eyes—to my interaction. The elementary particles we attempt to observe in isolation, like the electrons we fire at the slits mentioned in the previous chapter, lack that pre-established definition because they haven't yet established a relationship with anything else in the universe.

Classical mechanics has tried to convince us for a long time that things are defined and have inherent properties before anything else. However, we know very well that all matter is made up of the same fundamental particles, and it's the relationships established between those particles that cause them to define themselves and acquire "observable" properties before our eyes. The elements that make up my body—and yours, and everyone else's—are the same as those that make up the computer I'm writing on, the chair I'm sitting on, the cat glaring at me, and well, absolutely everything that exists. Where my fingers end and where the keyboard begins is simply a matter of order. If you could observe what happens at the subatomic level, you would really only see a sea of particles interacting with each other, and it would be quite difficult—not to say impossible—to distinguish which particles belong to the keyboard and which particles belong to me. At the scale of elementary particles, it would be like trying to figure out which drop of water belongs to a country and which is part of international waters. However, it's obvious that, from our scale, there's a clear distinction between my fingers, the keyboard, the other components of the computer, the cat, the chair, and that darn table. What creates that clear distinction? The order and the relationships, of course.

It's the way quarks are ordered and related that gives rise to protons or neutrons, depending on how they do it.

It's the way electrons, protons, and neutrons are ordered and related that gives rise to different atoms, whose properties are clearly defined and distinguished according to those relationships.

It's the way these atoms are related and ordered that gives rise to the molecules whose observable properties we know, based on how they relate.

It's the way these molecules are organized and related that gives rise to crystals and other substances whose properties we understand.

It's the way we order and relate all that matter that gives rise to objects and material structures, like the table we were discussing earlier.

In short, it's the relationship between its components that defines everything we know. It's the "observation"—the interaction—of its components with each other and with the rest of the surrounding universe that defines their characteristics.

What quantum mechanics has been telling us for many years is something as simple as this: anything that doesn't relate to anything else is still obviously undefined, because it's the relationship between things that makes observable, defined properties emerge.

# DISORDERING TIME II.

In the first part of this chapter, I mentioned that general relativity tells us that the concept of simultaneity is relative. And that the concept of "now" or even the order in which different events occur can be different for different observers. In reality, general relativity says much more than that about time, but I didn't want to overwhelm you at first.

General relativity unites space and time into a single entity known as spacetime. That's why it's often treated as a fourth dimension. But the important thing about this change isn't so much the fact that it fuses two previously independent entities, but that they also lost their universal and absolute nature in the process. The space and time of classical mechanics are universal and independent of the things they contain or that flow within them. According to classical mechanics, there's an empty space that we fill with things made of matter, energy, fields, forces, etc. And there's also a time that flows in a single direction and drags everything along with it, whether it wants to or not, all in that direction. General relativity shatters this mental image by constructing a single entity, spacetime, which is also totally dependent on and malleable by what it contains. It's still understood that it exists before everything else, but instead of being a fixed three-dimensional grid, indifferent to what it contains, it's entirely deformed and transformed by the presence of matter, energy, fields, etc. And in doing so, it also does away with the idea that gravity is a "force" and instead becomes simply the consequence of these deformations: the Earth doesn't pull you toward it with some magical invisible force; the Earth is deforming the spacetime it occupies in such a way that it makes you inevitably fall toward it. You can take a stretched sheet and place some weighted balls on it—what happens? The sheet deforms, and if you now place something on it, it will roll toward the ball. Just like that. But wait, we were talking about time.

The thing is, if you've been reading carefully, it's not just space that deforms. Space and time are one entity, so both things are deformed. Yes, time is deformed by the presence of masses. How? Well, the closer you are to a massive object, the slower time passes. In other words, you age less living at sea level than you would living in the mountains. And that effect is so real that the clocks on GPS satellites include corrections to account for that time dilation relative to Earth, to avoid their predictions being completely wrong. According to what we've learned from Einstein —and as we've already confirmed—time doesn't flow the same way throughout the universe, nor is it an entity separate from us: it governs us as much as we govern it. Since we are objects with mass (energy), it defines our path as much as we define it. So why do we insist on continuing to understand it as a superior entity to us? Why do we grant it that power over us? Perhaps it's time to redefine our relationship.

# MEMORIES OF SANTIAGO.

My relationship with studying physics has always been quite toxic. Like a love-hate relationship, studying physics has always felt like something I needed to do, while at the same time being a mental torment. This torment reached its peak while I was in my beloved Santiago de Compostela, after they explained to me, about twenty-five million times, the mechanism of symmetry breaking by which fundamental particles acquire mass—thus allowing them to interact gravitationally with each other. That's when I decided I didn't want a future dedicated to scientific research. Not because of the people who explained that mechanism to me, to be clear, but because my mind was already living in a constant state of abstraction that I could no longer sustain.

Looking once again toward the past—the only place you can look—I feel like I spent much of my adolescence and post-adolescence completely disconnected from the world. Or trying to disconnect from it. I suppose, consciously or unconsciously, I was so afraid that the world would define me through its observation that I tried to avoid interacting with it at all costs, to remain in my true and only state that feels real: superposition. However, perhaps precisely because I heard so much about the mechanism of symmetry breaking, walking under Santiago's rain and dancing under the lights of its nightclubs, I began to feel a certain energy (mass) emerging from me, forcing me to interact with the world.

## JAITZULF.

"Santiago. A quiet, gray place, with a clock from another time. There, I put my mind back together along with my solitude. I went back to doing some of those things others wouldn't understand for their lack of utility or logical sense, some of them in social settings. I spent a bit of time trying to fuse my solitude with partying, and while it often worked, the impact on my mind—feeling that I no longer had anything but one world—defined—was really intense.

I liked that talk about the gray a lot. I've always felt comfortable in the gray, in ambiguity. I've always avoided that constant duality that forces us to choose between black and white in every aspect of our lives. I've always been quite gray, but not because I'm quirky, but out of conviction. Defining and positioning oneself is fine, but the reality is gray; reality is full of nuances. And while I'm aware that very few people care about nuances, I'm afraid I'm one of them. I like to appreciate a good gradient, I like to see the full richness of the world, and I will never understand the need to choose just one part of it. I understand that it helps to keep focus, to follow the path. But the world wasn't created with marked paths—others made them.

Turning my worlds into just one—into one alone—led me to constantly and belatedly take part in all those battles I never wanted to fight. And that's how I reached my first real collapse: everything lost its charm. When I was with people, I felt alone. And when I was alone, I felt even lonelier. Nothing made sense anymore; everything felt indifferent. I listened to music, clinging to that silly idea that it would help me escape, but it was useless. Every song, no matter how silly or empty, brought thoughts into my head about all the things I was trying to avoid. Because that's what music does: it activates feelings, thoughts, ideas. And so, song by song, I fed my apathy. No song could understand me; no song was written to do that—just to accompany me. And somehow, reprimand me—damn it.

In Galicia, I lived with a woman, among others. This woman, with whom I had a good relationship, found out one drunken night that I was attracted to people of the same sex. From that moment on, to her, I was no longer myself, but a variant of myself in the form of a gay friend. It amused me one afternoon, but after that, I didn't want to participate in it anymore: I don't like it when people change their behavior toward me or assume that my tastes, aspirations, or determinations in life must one hundred percent conform to the stereotype they've created in their minds. Our relationship deteriorated from then on. I'll never forget how, on the last day of our time living together, while I was waiting to say goodbye, she left without even saying goodbye to me. From here, I apologize for having been myself and not the image you had created of what I should be."

Reading this now, I'm amazed by how differently I felt about that past when I wrote it compared to how I feel about it now. The farther I am from that past, the more it softens in my mind. Not in a fantastical or uncritical way, but in a way that allows me to appreciate its necessity in bringing me to where I am now. But also, I feel in a way that I can't quite explain, that that past doesn't even belong to me anymore. It's as if that past collapsed in on itself to give way to the great explosion of possibilities I'm living now, to give rise to all the chaos I can't stop loving.

# SINGULARITY.

The Big Bang, like Schrödinger's cat, is one of those ideas that has been misinterpreted to exhaustion over time—it's the nature of chaos.

The Big Bang theory is supposedly a physical theory meant to describe the origin of the universe, and yet all it really tells us is that the origin of the universe is a singularity. In other words, it has no idea what happened at the point of origin. All it does is describe everything that's happened since moments after the universe was created until today—which is no small feat. However, it's a theory that's become known as an answer to something it never actually answers.

But, but, what do you mean it doesn't? The Big Bang theory says that the entire universe was once concentrated at a single, infinitely dense point and then exploded, giving rise to everything we know. Right?

Well... no.

If we take the equations of general relativity and trace them back to the origin of the universe, what we find is a point where those equations stop making sense. Yes, one might say that mathematically it's an infinitely dense point, but physical reality cannot be infinitely anything. The singularity at the origin of the universe is a point where the physical reality we know ceases to exist—as it shouldn't surprise us—and, therefore, is indescribable by any physical theory. Surprise! Physics doesn't work until there's a physical world it can describe.

Surprisingly, the most fascinating and most misunderstood thing about the Big Bang isn't the fact that it can't describe the origin of the universe. The most fascinating thing is how the expansion —not explosion—from that singularity, from that point where the entire universe was concentrated, doesn't mean there's a single localized point in the universe that exploded and caused space to expand around it, but rather that every point in the current universe is that same point of origin that hasn't stopped expanding and creating new spacetime within itself. The Big Bang happened in every point of the universe. In other words, what's really fascinating is that, if the Big Bang theory is correct, the entire universe is nothing more than a point in disorder.

# CUPID'S ARROW.

In Roman mythology, there's a winged being who carries a bow and a magical arrow that, if it hits you, makes you feel a tremendous sense of love. I'm talking about Cupid, obviously. Cupid symbolizes the sudden act of falling in love that humans often experience, and his wings represent love's fleeting and ever-changing nature.

Sometimes I think that existence itself was struck by Cupid's arrow, and that's precisely what gave rise to the beginning of time, if such a thing existed. One day, nothing—or everything bunched up and still, who knows what—and suddenly, BANG, a big BANG, a very BANG, an enormous BANG. We say BANG because we can't really understand what, but something happened, and suddenly everything started interacting with everything else around it, exchanging energy, information, transforming through interaction with its environment, disordering. Suddenly, everything started sharing its love in a frantic, unstoppable way, as if Cupid himself had shot the arrow of time.

# AND LOVE.

JAITZULF.

"Love doesn't exist,
it's just friendship with things."
— said someone I hardly recognize anymore.

# MEMORIES OF THE BIG BANG.

Luckily for us, photons don't have the consciousness to feel the passage of time, so everything I mentioned in the previous chapter can be put in our "magical phenomena of the universe" drawer, and we can move on with our lives.
Well, since we've opened the drawer…

Shall we talk about quantum superposition again? A few chapters back, I mentioned this phenomenon when discussing Schrödinger's infamous cat, but I'd like to delve into it in more depth, without analogies.

Besides the double-slit experiment, many other experiments have been conducted to try to understand the phenomenon of quantum superposition. Most of them consist simply of measuring the binary properties—which only admit two opposing results—of elementary particles, such as spin.

Spin is a property of electrons that can be crudely and inaccurately (but sufficiently) interpreted as the direction in which they rotate around themselves (like the Earth, for example). When we measure them, we find that some electrons spin to the left, and others to the right. And there are no other options. All the electrons observed and measured in the universe spin one way or the other. We'll call these two states from now on positive spin and negative spin.

The thing is, if you design a machine capable of detecting electron spin and fire electrons (whose properties we don't know) through it, the result you get is that half the electrons in the world have positive spin and the other half have negative spin. So far, nothing strange. But if instead of measuring a single property like spin, we want to measure two different properties of the electron, things start to get a little weird. Let's imagine we want to measure another binary property of electrons that we'll call color (let's suppose, for argument's sake, that half the electrons in the world are pink and the other half are blue). What you'd expect, if spin and color were two unrelated properties, is that if after measuring the color of the electrons and separating them into pink and blue, we measure their spin again, both the pink and blue electrons would still be half positive and half negative, right? Well, they are. But. But it turns out that, if after doing this, we set up a third machine to measure the electrons 'color again, logic tells us that the ones we initially separated as pink will remain pink, and the ones we separated as blue will remain blue, right? Well, no, not this time. It turns out that after measuring their spin, half of those that were initially pink are now blue. And vice versa. This might make us think that somehow, we're changing the electrons 'color when we measure their spin. Well. It turns out that if we design a machine to separate electrons into pink and blue and then direct them all to a detector that measures their spin, if we fire only electrons with negative spin, we'd expect, based on the previous experiment, that the machine detecting their color would affect their spin, once again giving us half positive and half negative electrons. But no. It turns out that if we fire negative-spin electrons into that machine, they all remain negative when they exit. And if we fire only positive-spin electrons, they remain positive. Measuring color doesn't change spin in this case! And. And not only that, but things get even weirder. Much weirder. It turns out that if we place a small barrier in our machine, blocking the path of electrons after separating them by one of the two routes (blocking the path of either pink or blue electrons), given what we saw in the previous experiment, we'd expect only half to get through, but all of them would still be negative, right? Well, no! Suddenly, they're half positive and half negative again. It turns out that, for some reason, if you only measure pink or blue electrons separately, half of each will have positive spin or negative spin, regardless of their original spin. But if you measure pink and blue simultaneously, if they were all negative, they'll remain negative, and if they were all positive, they'll remain positive. What a mess with these electrons!

Let's try to organize our thoughts: if when measured separately, the electrons that come out blue are half positive and half negative, and the pink ones are the same, that means that when we measure them simultaneously, the electrons cannot be passing through either the blue or pink path because then they wouldn't be able to maintain their spin. If they didn't pass through the machine via either path, that leads us to think that either they didn't pass through at all (but then we wouldn't have detected them at the end), or they must have passed through both simultaneously (something we've also never measured). Since we don't exactly understand what electrons do in situations like this, we say they're in quantum superposition during the experiment and collapse into a defined state only when they're observed (measured) at the end. But, despite not fully understanding what happens, we have a whole theory and mathematics developed that fits perfectly with both the before and after of interacting with those magical and stupid electrons.

The Copenhagen interpretation, the most widespread, proposes that we understand the electron can indeed pass through both paths at once. That is, we interpret that positive-spin (or negative-spin) electrons are both pink and blue at the same time, allowing them to pass through the machine via both paths, thereby arriving intact at the end. Whereas when we block one of the two paths, we force them to be pink or blue—we make them collapse into one of the two states—which alters their spin.

In any case, even though we don't know what electrons do during this experiment, it's clear and undeniable that it perfectly illustrates the fact that, as indicated by the uncertainty principle, there are properties that are impossible to measure accurately at the same time. In the case of our experiment, the properties are color and spin. In practice, the position and momentum (mass times velocity) of a particle are impossible to determine precisely at once (measuring one alters the other) or—and this one interests me more—energy and time.

Well, we'd better close the drawer of little magical things again for a while.

# QUANTUM SUPERPOSITION.

Luckily for us, photons don't have the consciousness to feel the passage of time, so everything I mentioned in the previous chapter can be put in our "magical phenomena of the universe" drawer, and we can move on with our lives.
Well, since we've opened the drawer…

Shall we talk about quantum superposition again? A few chapters back, I mentioned this phenomenon when discussing Schrödinger's infamous cat, but I'd like to delve into it in more depth, without analogies.

Besides the double-slit experiment, many other experiments have been conducted to try to understand the phenomenon of quantum superposition. Most of them consist simply of measuring the binary properties—which only admit two opposing results—of elementary particles, such as spin.

Spin is a property of electrons that can be crudely and inaccurately (but sufficiently) interpreted as the direction in which they rotate around themselves (like the Earth, for example). When we measure them, we find that some electrons spin to the left, and others to the right. And there are no other options. All the electrons observed and measured in the universe spin one way or the other. We'll call these two states from now on positive spin and negative spin.

The thing is, if you design a machine capable of detecting electron spin and fire electrons (whose properties we don't know) through it, the result you get is that half the electrons in the world have positive spin and the other half have negative spin. So far, nothing strange. But if instead of measuring a single property like spin, we want to measure two different properties of the electron, things start to get a little weird. Let's imagine we want to measure another binary property of electrons that we'll call color (let's suppose, for argument's sake, that half the electrons in the world are pink and the other half are blue). What you'd expect, if spin and color were two unrelated properties, is that if after measuring the color of the electrons and separating them into pink and blue, we measure their spin again, both the pink and blue electrons would still be half positive and half negative, right? Well, they are. But. But it turns out that, if after doing this, we set up a third machine to measure the electrons 'color again, logic tells us that the ones we initially separated as pink will remain pink, and the ones we separated as blue will remain blue, right? Well, no, not this time. It turns out that after measuring their spin, half of those that were initially pink are now blue. And vice versa. This might make us think that somehow, we're changing the electrons 'color when we measure their spin. Well. It turns out that if we design a machine to separate electrons into pink and blue and then direct them all to a detector that measures their spin, if we fire only electrons with negative spin, we'd expect, based on the previous experiment, that the machine detecting their color would affect their spin, once again giving us half positive and half negative electrons. But no. It turns out that if we fire negative-spin electrons into that machine, they all remain negative when they exit. And if we fire only positive-spin electrons, they remain positive. Measuring color doesn't change spin in this case! And. And not only that, but things get even weirder. Much weirder. It turns out that if we place a small barrier in our machine, blocking the path of electrons after separating them by one of the two routes (blocking the path of either pink or blue electrons), given what we saw in the previous experiment, we'd expect only half to get through, but all of them would still be negative, right? Well, no! Suddenly, they're half positive and half negative again. It turns out that, for some reason, if you only measure pink or blue electrons separately, half of each will have positive spin or negative spin, regardless of their original spin. But if you measure pink and blue simultaneously, if they were all negative, they'll remain negative, and if they were all positive, they'll remain positive. What a mess with these electrons!

Let's try to organize our thoughts: if when measured separately, the electrons that come out blue are half positive and half negative, and the pink ones are the same, that means that when we measure them simultaneously, the electrons cannot be passing through either the blue or pink path because then they wouldn't be able to maintain their spin. If they didn't pass through the machine via either path, that leads us to think that either they didn't pass through at all (but then we wouldn't have detected them at the end), or they must have passed through both simultaneously (something we've also never measured). Since we don't exactly understand what electrons do in situations like this, we say they're in quantum superposition during the experiment and collapse into a defined state only when they're observed (measured) at the end. But, despite not fully understanding what happens, we have a whole theory and mathematics developed that fits perfectly with both the before and after of interacting with those magical and stupid electrons.

The Copenhagen interpretation, the most widespread, proposes that we understand the electron can indeed pass through both paths at once. That is, we interpret that positive-spin (or negative-spin) electrons are both pink and blue at the same time, allowing them to pass through the machine via both paths, thereby arriving intact at the end. Whereas when we block one of the two paths, we force them to be pink or blue—we make them collapse into one of the two states—which alters their spin.

In any case, even though we don't know what electrons do during this experiment, it's clear and undeniable that it perfectly illustrates the fact that, as indicated by the uncertainty principle, there are properties that are impossible to measure accurately at the same time. In the case of our experiment, the properties are color and spin. In practice, the position and momentum (mass times velocity) of a particle are impossible to determine precisely at once (measuring one alters the other) or—and this one interests me more—energy and time.

Well, we'd better close the drawer of little magical things again for a while.

# EVERYTHING IS RELATIVE.

One of the great critics of quantum theory—or more precisely, the Copenhagen interpretation—at the time was Einstein. Einstein, who contributed (though to a lesser extent than others like Planck or Bohr) to the idea of quanta—the idea that, at a fundamental level, particles take on finite, concrete values rather than continuous ones to define their properties, like energy, which is the idea behind quantum theory—was highly critical of the idea that particles could be in simultaneous states, collapsing randomly, and the consequences this had for our understanding of the world. Curiously, while he criticized the idea that determinism might come to an end due to the randomness implied by quantum mechanics, he introduced ideas in his general theory of relativity that seemed to break with absolute truths as well.

Einstein, while saying "God does not play dice," tried to get the world to understand that time does not pass the same way for everyone. He tried to get the world to understand that time passes so differently for some compared to others that even two twin brothers, if placed in extreme situations, could end up not being the same age. I'm talking about the twin paradox. Oh, looks like the drawer's opened again.

The twin paradox suggests that two twin brothers born on Earth —on the same day, place, and time, of course—are separated at birth, with one sent into space in a ship that reaches speeds close to the speed of light. When the spacefaring twin returns from his journey, he finds that his brother is now much older than he is (how much older depends on how long the journey lasted). This paradox arises from the same phenomenon that causes light not to experience the passage of time: time dilation.

What's particularly curious about this paradox is that during the twin's journey, if instead of observing from Earth, we do so from the spaceship, the spacefaring twin does not perceive time to be passing more slowly for him. And, in fact, during certain parts of the journey, he would see that time on Earth is flowing more slowly. The truth is, in the end, the twin who will have experienced less time overall will always be the one who traveled on the ship. Why? Because he breaks a symmetry—as always. The traveling twin accelerates to enter and leave Earth, and that acceleration—which only one of them experiences—breaks the symmetry between the twins, causing the spacefaring twin to experience less net time than the Earth-bound twin.

This paradox is a great way to understand the concept of relativity in Einstein's special relativity: that time passes relatively differently for each entity, with each having its own clock, its own time, meaning that two people born at the same instant can end up with vastly different ages does not, under any circumstances, imply that each person has a different reality. In this entire paradox, there is only one possible reality: the spacefaring twin will be younger than the Earth-bound twin when they reunite. And this will be true for both of them. Always. There is only one reality. The relativity in Einstein's theories never leads to accepting different realities, but rather how each entity experiences and is affected by the same facts.
In other words, while the phrase "everything is relative" is often

used socially to settle disagreements and suggest that different perspectives are equally true—even if they don't align—Einstein made great efforts to conclude the opposite. In fact, Einstein's special relativity introduced—paradoxically—one of the very few things that don't seem to be relative at all: one of the fundamental constants of the universe. The speed of light, imposed as a limit by Einstein's theory—and it seems quite clear, also by reality—is one of the fundamental constants of nature, one of the few universal values that don't depend on absolutely anything else.

# MEMORIES OF GALWAY.

Sometimes it's hard to believe that I only spent two months in Galway. The sheer number of things I lived and felt, the number of people I met, and the number of moments that shaped my personality and worldview ever since, all happened in just two months—it's overwhelming. Everything went by so fast—and even faster when I was living through it—but, somehow, my mind has never been able to process how brief that time was. It's as if it had all been part of some intergalactic journey at the speed of light, and time had dilated.

"Well, I'm losing my train of thought: there I was, 25 years old, living abroad alone for the first time. I decided to settle in Galway, the most wonderful city I've ever walked. A city where, even if it was cold or raining, I always felt like going for a walk.

In Galway, I learned about many forms of love. I learned what it means to love your homeland; I learned what it means to love the unknown—and strangers; I learned what it means to truly love in an ephemeral way, and above all, I learned that all forms of love are equally important. Love, like everything, is beautiful regardless of its nature. It's beautiful in the eternity of its perfection—in the eternity of its circle—and it's also beautiful when it strays from that, when it becomes instantaneous and even shatters into pieces. The love of fragile, clumsy, human beings with limited durability is as beautiful as the eternal and sweetened memory of it. Love is always beautiful.

*Love is as beautiful when it's light as when it's snow.*

*In Ireland, I remember going to an overwhelming number of bars and ordering a Galway Hooker. And two. And three. And four. I also remember ordering whiskey—and being shocked by the prices. I remember dancing to songs I never thought I'd hear in a pub. I remember meeting people as wonderful as they were awful. I remember loving a lot, all the time. Loving more than three guys in the same week, loving in many houses, bathrooms, nightclubs, and streets. Loving by day, but especially by night. Loving the watery coffee from the stand 2km from home, where I walked every morning. Loving the cold wind by the shore. Loving that damn main street that looks like something out of a mythological tale. Loving those who pulled strings to get me a job, and those who got down to business. Loving those I only saw one night and those who wanted to include me in their routines. Loving a woman with just a "hello." Loving the sun every time it came out. Loving—more—the rain. Loving tears as much as laughter. Loving the where, the when, the what, and the who, regardless of the why. But, above all, I remember loving you. You who unsettled me every time you spoke to me—or every time you ignored me. You, who opened your home to me when I was left out on the street. You, who smiled with your whole jaw. You, whom I never heard from again. You, the only man I ever felt I could truly reach out and touch."*

# THE OTHER HALF.

Human interaction has long been simplified, just like almost every way we interpret the world, into a search for opposites that attract. As if the whole world were composed only of magnets and nothing else. There's also a great obsession with always wanting to complete everything, to create opposites for everything. These two ideas together have given rise to the grand narrative known as the search for the "other half."
The search for the other half—finding the partner who completes you—is a narrative that has not only infiltrated romantic stories. It's also, to a great extent, the story of physics.

From its beginnings, physics has been an observer of balance. Organizing and recognizing patterns isn't only the origin of time, as I discussed earlier while watching the sea, but it's also the origin of all knowledge. We understand everything as balance. We understand everything as symmetry. We understand everything as tending toward order. For example, when a ball falls from a roof to the ground, we interpret it as its trajectory toward energy equilibrium. The movement of pendulums, for instance, is understood as oscillations around a point of equilibrium that the pendulum constantly strives to reach. We interpret everything in terms of balance. This way of understanding the world inevitably requires every part of the universe to have a counterpart—equilibrium demands that every positive charge must have its negative equal, for example. Surprisingly, and often unacknowledged, forcing this view on the world has led us to become aware of the opposite: the world only exists because it is out of balance, because it is disordered, because it breaks the symmetries we try to impose on it. Because it is imperfect.

In an earlier chapter, I mentioned how this has been one of the most obsessive ideas I've had since I started studying physics: the notion that everything is the result of broken symmetries. The idea that the universe exists because it breaks the perfect mold we try to fit it into. Time is probably the most obvious broken symmetry: it only flows in one direction, always toward the disorder of the world, never toward order. But it's far from the only one.

The amount of matter and antimatter (particles with opposite charges to the fundamental ones we know) is remarkably asymmetric in the world, even though there is no apparent reason for it to be so, and despite all the physical laws leading us to expect otherwise. According to these laws, every particle should be the other half of its "other half," its corresponding antiparticle. And while we do observe antiparticles, which are stable and constantly generated in particle physics labs—they are an observed reality, not just a theoretical prediction—it is extremely rare to find them outside of those labs in our universe. Matter and antimatter annihilate each other when they interact, turning into pure energy. This means that if there were equal amounts —or even nearly equal amounts—of both, the world would have annihilated itself moments after it began. In fact, that's precisely what happens in the quantum vacuum and why it appears to be nothing. If we've come to exist, if anything observable in the universe is more than just emptiness, it's precisely because the amount of matter is far greater than the amount of antimatter, which allowed matter to form structures as it became disordered. If the world followed equilibrium, it would be completely empty.

Furthermore.

Matter, according to quantum field theory—which defines those elementary particles and antiparticles—wouldn't have mass if it weren't for a broken symmetry that causes particles to interact with the Higgs field. In other words, they wouldn't have mass if a fundamental symmetry of the interactions governing physical reality weren't broken.

And then there's.

Then there's chirality. Chirality is the property of an object that cannot be superimposed onto its mirror image. Think of a spring. If the spring coils to the right, its mirror image would coil to the left, and if you tried to superimpose them, it would be impossible. In this case, we'd say the spring has two chiral forms. In the case of the spring, both chiral forms are functionally identical: regardless of the direction in which the metal coils, it will compress and stretch the same. But the case of the spring isn't always true. When synthesizing certain molecules, chirality becomes extremely important. In the 1950s, a drug called thalidomide was developed to treat nausea in pregnant women. During its synthesis, the chirality of the molecule wasn't considered, resulting in many newborns suffering from malformations. The same substance, made with the same elements arranged in the same "order"—that is, the same molecule—could have such different consequences as alleviating the nausea of pregnant women without side effects or causing significant malformations in the fetus they were carrying. Chirality is yet another property that reminds us of the importance of gender equality in all fields (and of broken symmetry, once again).

In any of the above cases, what's important and meaningful—for better or for worse—is what breaks from expectations. It's what falls outside the norm. Luckily, the world isn't made up of other halves searching for each other to interact, because if that were the case, nothing would exist. It's neither perfectly symmetrical, nor does time flow in both directions, nor does it look the same when placed in front of a mirror. The world only exists and forms a whole because it is made up of interactions between oranges as complete as they are imperfect. Oranges whose definition is based on the potential of their interactions and their relationship to the world. The world only exists because, somehow, imperfect, asymmetric, broken things emerged. Stones capable of movement came into existence.

# THE POTENTIAL OF STONES.

Stones, as part of inert matter—being lifeless entities—have only one path in their existence: to seek equilibrium. When a stone breaks loose from a cliff, it's following its path toward energy equilibrium, falling to the point where its energy, relative to its surroundings, is lowest. Until the moment it detaches, the stone stays in balance with the other stones that form the cliff. But when—whether due to erosion, infiltrated water, or some other factor (such as a human deciding to build a tunnel nearby)—that equilibrium is broken, the stone has no choice but to seek a new balance. It must find a new place in the world where it can settle into equilibrium with its environment. And so, moment by moment, day by day, year by year, decade by decade, and millennium by millennium, stones do nothing but this: moving from one state of balance to another. Falling and rolling between equilibriums.

From now on, understand that when I talk about stones, I'm not referring only to what we commonly call stones but to any component of the world that belongs to inert matter. This, though it may cause some confusion, includes the very components that form living beings when isolated from them. In other words, the electrons that make up my body are just as much stones as the bricks that form my house or the iron atoms supporting the desk where the computer I'm writing on now sits. A stone is anything whose only potential, in and of itself, is to move from one state of balance to another, without even making a conscious decision. And despite this, physics is intent on understanding this wonderful world we live in as a collection of isolated stones. A collection of stones that, in brief moments, exert forces on one another in a cold and calculated manner.

Personally, I refuse to see the world in this way.

# MEMORIES OF GORRONDATXE.

Since the summer of 2020, Gorrondatxe beach, despite being full of stones, has become my favorite spot to go and read. In that context of the pandemic, being an isolated stone in search of equilibrium was the most my mind could aspire to. Gorrondatxe beach has witnessed all my recent "realization moments." All those moments when your mind clicks, and your perspective on the world shifts in some way. During my walks and reading sessions on that beach the following year, I remembered how important it had always been for me to move stones rather than be one. Although that same summer, I regained my love for reading and my ability to focus—perhaps because reality was so overwhelming, and I needed anything to escape from it—it wasn't until 2022 that I returned to the habit of studying physics. And by pure chance. While searching for books to read at my favorite bookstore, I came across a book titled Helgoland by Carlo Rovelli, with a cover that fascinated me. I read it that summer and shed a tear when I finished it—one of those tears that fall when you realize a wall in your world has crumbled, when you become aware that your perception of everything around you will never be the same. After a year of mulling over the ideas the book sparked in my head and trying to discuss them—fervently, as reality should be debated—whenever life gave me the chance, I reread the book in the summer of 2023. That's when my mind began to piece together all the chaotic ideas it had generated.

That's when I finally decided to give shape to this text.

Perhaps influenced greatly by my personal experiences, it has always baffled me how physics is mostly interpreted as the study of isolated entities and absolute definitions (or at least as absolute as possible). It's curious how a science that requires its first step to be choosing a reference point in relation to which everything else makes sense, then attempts to deny its relational significance. Every number, every definition, every piece of knowledge, everything we know, predict, calculate, and describe is always done in relation to something else. "25 what?" I often ask my students in class. "Units! If you don't write the units, this result means nothing!" Yes, I'm that teacher. But that's exactly where the key to what I'm talking about lies. When we calculate things, when we assign them a numerical value, it's always relative to something. Talking about velocity, position, momentum, time, pressure, temperature—anything, of any object or being—requires doing so in relation to something else. Saying you're 32 means nothing. Saying you're 32 years old means that, since you were born, the Earth has made 32 orbits around the Sun. In other words, defining your age depends on the Earth. But more than that, to define what a year is, the Earth needs the Sun —and the Sun needs the Earth. The very unit of a year is, in itself, a relationship between two things. The same is true for any other unit. For example, a meter is a specific relationship (of distance) between two points. And so it goes with all units. This isn't just about units—physical laws themselves, defined by equations, are nothing more than relationships between concepts, as we saw in the previous chapter! Relationships between quantities that are, in many cases, already defined as relationships themselves. And yet, we often pretend otherwise. Classical physics—and its earliest interpretations—left us with such a well-ordered, absolute, incredibly functional world, presented as a perfect place defined by absolutes, that we believed it was true. We believed it so much that, when we eventually advanced in our exploration of the world's fundamental characteristics, when we began interacting with the smallest, most essential aspects of the world, we had already forgotten this.

The reality is that you don't need to reach quantum mechanics to interpret physics as a set of relationships. You can do it with one of the earliest concepts defined in physics. Defined, that is, only in a manner of speaking, since, like time, there's no clear definition of mass. And its meaning isn't even consistent across different physical theories. Mass was originally—and commonly—is understood as the amount of matter in a physical object. However, this definition was deconstructed by modern physics a long time ago. In general relativity, mass is essentially equivalent to a particle's energy (its capacity to transform). And in quantum field theory, as already mentioned, mass isn't even necessary to make the Standard Model of particles work and define them. Instead, mass appears only as a consequence of the interaction of particles with an external field (the Higgs field). We don't really know what mass is—at least, not by itself. Could it be that it's nothing on its own?

Mass is understood in its earliest conception through the following equation:

$$F = ma.$$

Force equals mass times acceleration.

This equation is usually interpreted to mean that the force applied to an object equals its mass times its acceleration. In other words, the movement of a mass accelerates according to the force applied to it. In this interpretation, mass is taken for granted as the intrinsic property that defines the object. But if we follow this equation, mass is nothing more than the measure of the relationship between two observed phenomena: the object's acceleration and the force applied to it. Deciding that mass is an intrinsic property of the object is, in reality, an interpretive decision. From its origin, mass has really only ever been a relationship.

Let's now consider the next equation:

$$E = mc^2.$$

Energy equals mass times the speed of light squared.

This equation reveals that mass is essentially the same as energy since it's related to energy by a constant number. Talking about an object's mass or its energy gives us fundamentally the same information about the object. (If one thing is defined by a number and another by twice that number, the second doesn't tell me anything new; I can obtain it by simply multiplying the first by two.) But if we make the effort to reinterpret this equation, another possible reading emerges: mass is the relationship between an object's energy (its capacity to transform) and the speed limit of the universe (the speed of light in a vacuum). In other words, mass is here—the way we might choose to interpret it—the relationship between a particle's abilities and the limits imposed on it.

In any case, the reinterpretation of mass described here is just one example of many concepts that show that physics, despite what we've been led to believe, has always been open to being understood as relationships between observable phenomena, rather than as the study of objects 'intrinsic properties. Granting such quantities—like mass—the status of being definitive was merely a choice made by those who constructed this way of understanding the world.

# QUANTUM ENTANGLEMENT.

One of the most surprising and fascinating quantum phenomena is entanglement.

According to the Copenhagen interpretation—the most widespread and well-known interpretation of quantum mechanics—superposition means that quantum particles exist in more than one state simultaneously until observed. This implies that if we take a pair of particles that start in the same related state, separate them, and send them in opposite directions to very distant locations, as long as they don't interact with anything else (i.e., are not observed by anything), they should remain in a superposition state. If we then measure one of them, collapsing it into a defined state, the other should remain in superposition until it is also observed. In other words, knowing the result of the first measurement should not give us any information about the result of the second measurement. However, what we actually observe is that both results are instantly correlated, as if the particles were telepathically communicating how they should collapse as soon as the other one does, as though there were a "spooky action at a distance." You might initially think, as most people do, that this result occurs because the particles already had predefined characteristics when they were launched, not in that magical superposition state that crazy physicists try to convince us of. However, this experiment has been modified and repeated in various ways, such as changing how one particle is measured but not the other. The results from these experiments have unequivocally led to the conclusion that there is a much stronger correlation between these particles than would be expected if the effect were due to our ignorance of some already predefined parameters.

This experiment posed a great dilemma for the scientific community, particularly for Einstein, because it seemed to suggest that a fundamental principle in physics—one that Einstein himself helped solidify—was being violated: locality. Locality is the principle that I described in the first chapter of this book. It states that all information, interactions, relationships, energy, objects, mass, and physical entities, in general, have a speed limit: the speed of light in a vacuum. Locality tells us that things can only influence and exchange information over a certain amount of time, at most, with those at a distance that light can travel within that time. In other words, nothing is instantaneous, and whatever is happening 150 light-years away from you right now will never affect you because, in 150 years—the earliest time it could reach you—you will likely be dead. Anyway, we were talking about entanglement. In this phenomenon, it appears that information travels instantly between the particle we measure first and the other one. Only this would explain why they act in a correlated manner. The general interpretation's way of resolving this is that no real information travels between the particles. Since the collapse of the first particle's measurement is a random event (there's no way of knowing beforehand what the specific result will be), it means that even though the second particle will behave in accordance with the first, this phenomenon can never be used to transmit information. The information sent is random, and thus tells us nothing.

My goodness, what a mess! To the drawer with it!

# ENTANGLEMENT.

It often happens that people who have lived together—sharing many life experiences—and have deeply gotten to know one another's actions, ways of living, loving, feeling, gesturing, talking, and being, are the very ones who, after growing apart for a time, become less able to recognize or understand each other when they reconnect.

Growing alongside someone, living with them, sharing anything, means interacting with them. Interacting, observing, measuring, relating. People who form a life together inevitably define themselves based on each other. Relative to one another. This often leads to the situation where, when their paths diverge and the interactions and relationships start to happen with others, two people who were once entangled, whose destinies seemed linked and bound to one another, are no longer connected to the point where there's no longer any correlation between them. To the point where, when they meet again, they see each other as complete strangers. That entanglement, that correlation between what one did and how the other responded, was conditional on not interacting with anything else—or at least not with enough intensity to influence them significantly. It is in our relationships that we recognize, define, and find ourselves.

Even though it may not seem like it based on how we structure our discourse or the obsession that permeates society with identity, one of humanity's greatest successes has been exploiting this fact to build societies and establish relationships that allow us to rule the world as we please. The success of societies has always rested on establishing an order over human relationships, on governing interactions while giving each individual their own clock, their own sense of self.

# MARRIAGE.

Marriage is probably the greatest proof that, consciously or unconsciously, we've always known as a society that what matters is defining relationships, not individuals. Marriage is also undoubtedly the fundamental pillar that has sustained the creation of the societies we live in. The institution of marriage is a way to order individuals into entangled pairs whose definitions cease to be individual and become relative to the world. It's a social mechanism to ensure order, a non-chaotic interpretation of relationships, in our society.

Although this meaning may be fading now, we shouldn't lose sight of the fact that marriage originally arose to unite individuals of opposing properties (man-woman) and assign very distinct, rigid roles to each. Marriage was created to give all power to one over the other. It emerged as a form of quantum entanglement: the definition of one automatically and involuntarily collapses and defines the other, at any time and in any place. And for far too long, society has interpreted this phenomenon just as it does with entanglement: as a ghostly action at a distance.

Marriage has been a social mechanism that allows for the orderly raising of new workers who obediently follow the passage of time in an incredibly efficient way, but it's also a space that has caused immense pain and suffering for far too many women. It has also been a space that prevented half of society from living in superposition and exploring their ocean of possibilities.

## A STORY ABOUT THE VOID.

One story I tell every year to my high school students is the story of Mileva Maric. The most important story in modern physics. Mileva Maric studied mathematics and physics at the Polytechnic School of Zurich, where she excelled. But despite being a brilliant mathematician, her name never entered the history of science as such. Instead, it became known as that of Einstein's first wife and the mother of his children. Einstein and Mileva Maric became entangled while studying together, but with the arrival of their children, he continued his university studies while she had to abandon hers. The responsibility of raising the children, as well as the demanding conditions Einstein imposed on her as his wife, forced Mileva Maric, like many other women, to give up her promising career as a scientist. After years of marriage, Mileva Maric eventually separated from the genius Einstein because she could no longer bear his demands as a husband. This same husband, it seems, once declared: "I need my wife; she solves all my mathematical problems." Years later, when the couple's letters were made public, it was revealed that Einstein referred to the theory of relativity when speaking to his wife as "our theory of the relativity of motion."

The story of Mileva Maric is not just the story of Mileva Maric. It's the story of humanity.

# WORK.

Information. Information is the key to our entire world. In what we call the era of globalization, everything boils down to sharing information. We live connected 24/7, constantly linked to one another. We've filled the world with devices connected to an information exchange network called the internet, and we dedicate ourselves, at every moment, to feeding that network with more and more information each day. While we believe we've invented something wonderful, in reality, we haven't invented anything—we've merely imitated reality.

Reality is a network of information. Or rather, a network of information exchange. Reality is a place where particles interact constantly and unceasingly with all others within reach, exchanging information among them. They exchange information through the fundamental electromagnetic, weak, and strong interactions via boson exchanges (virtual particles), and they also do so through their position and location in space (gravity). Everything, at every moment. As if they couldn't do anything else. Interact, interact, and interact. Relate, relate, and relate.

Many pages ago, I explained the concept of entropy and how it's associated with disorder. To say that a system is more disordered is the same as saying we need more information to fully understand it. Entropy and information are, therefore, also linked. Entropy, which measures a system's disorder and—

remember—always increases, is also a measure of the information needed to understand a system. This leads to an inevitable conclusion: every second that passes, we need more information to understand the world. Time passing, the world becoming more disordered, means that it becomes more complex and richer in information. If we accept the Big Bang as the correct theory, this narrative fits perfectly: the universe initially was a single point where everything was concentrated and was one single thing, so the information needed to understand the universe in that state was minimal—we only needed to know a single point. As interactions occurred and relationships between parts of the universe were forged, it became increasingly complex, requiring more and more information to comprehend. It's like a Rubik's cube, which starts in a perfectly ordered and easily understandable state but becomes more disordered and harder to understand with each interaction. The flow of time, in this sense, is the creation of new information accessible to us. Requiring more information to describe the world is simply the same as saying that the world contains increasingly more information. Information generated through fundamental interactions, through the energy exchanges of the particles that make up the world. Information generated through work.

In classical physics, a system is said to perform work when it transfers energy to another (exerting force) and causes changes (typically displacements) in it. This definition closely mirrors the social understanding of work before the concept of work came to mean selling one's lifetime for money. It was defined this way during the era of brute labor, when forces, Newtons, and non-intelligent machines ruled the world. In modern physics, it's a concept that has almost disappeared. Energy transfers are still discussed, but as we've moved away from the concept of force and toward the concept of interaction, these are now understood purely as energy exchanges without further conceptualization.

The concept of work is once again closely tied to entropy. Entropy, that measure of a system's disorder, can also be understood as "the amount of energy unavailable to do work." If entropy grows systematically and inevitably, and if this is the amount of energy that, in a thermodynamic process, is unavailable to do work, we could infer that the amount of work the universe can do on itself is constantly decreasing. In other words, the amount of available energy the universe has to change itself is shrinking. As if the universe were a living being, the older it gets, the more it ages, and the less capacity it has to work. As the universe becomes more disordered, it also becomes less capable of making changes to itself. In energy terms, this is easy to understand: as the universe becomes more disordered, its energy becomes increasingly evenly distributed (it's no longer concentrated in specific points), making energy exchanges—or interactions between things—more difficult. The potential moment when the universe reaches its most disordered state and becomes unable to make any more changes—unable to transform or allow its components to interact—is known as the heat death of the universe. For humans, ceasing to work is also understood as social death, as the end of their time.

Doing work, generating information, exchanging energy, letting time pass, or disordering the world are, in essence, the same concept. Trying to order the world is fighting against the inevitable.

# DISORDERING THE WORLD II.

Loving is, without a doubt, the ability to disorder the world for another being. Love, that thing I've already told you I don't know, is the most fundamental symmetry-breaking tool.

Love effortlessly breaks time.

"Suddenly, your gaze meets that person's, and the world stops for you for a few seconds. Immediately after, your whole body speeds up, as if it had to catch up with the rest of the universe that kept changing, that kept flowing in time. They smile at you, and you feel that pause again, followed by its corresponding acceleration. It's 6 in the afternoon, it's dark outside, it's winter, and you're studying in the library. None of that matters to you. The only disorder you care about is the one they are causing in you."

Love breaks charge symmetry.

"There was a time in human history when a bunch of people spent their time observing whether male penguins properly fuck each other or fall in love, only to then argue whether it was acceptable for two men to do the same. Doesn't it sound pathetic just to think about it? Imagine wanting to understand the world as a place where the ability of beings to love and relate to one another is understood as a need to match each positive charge with its

opposite negative —to annihilate— instead of simply enjoying the disorder they are causing in you."
Love breaks symmetries by generating mass (energy) and warping the world.

"If there's one thing I've learned as a high school teacher, it's that there is no greater source of energy than adolescence. No one loves the world more than teenagers. They are the force that causes the most disorder. Youth, eternally criticized for being irreverent, for not accepting authority, the laws of the adult world, and erasing the past of those who age, is the condition of possibility for everything that is to come. They are the ones who must inevitably shape space-time for the rest of us. To deny youth, to ridicule it, is to go against the necessary disorder that allows the world to continue existing."

Love is everything that allows existence to make sense, not in a romantic sense, but because it is simply the most radical act of rupture: love is the singularity of the Big Bang, it is breaking with freedom.

# MEMORIES OF COVADONGA.

*"If there's one thing I've realized while traveling, it's that what I enjoy most is the journey. Not because of the excitement about what I'll find when I get there—although that's part of it—but because I feel freer than ever. When you're on a train, plane, bus, or driving a car, you're free to just be and observe. You can't, and shouldn't, do anything else. Exist and allow existence through observation. There's no room for stress or the incessant feeling that you must always be doing something. And that's wonderful. Delight in existence. Wow. It's the best thing in life, especially if you feel like you're in a little bit of danger at the same time. It's incredible but true: it's in the places where I appear least free that I feel most free.*

*Enjoy existing,*
*it could all end*
*at any*
*moment."*

# MUSICAL CHAIRS.

In recent years, much has been discussed about freedom. Years, if not decades or centuries—I don't know. It doesn't really matter. I suppose freedom has been a topic that has preoccupied humanity throughout its history.

Since the development of a productive mindset, driven by the capitalist system and its need for infinite (infinite) growth—the same mindset that disordered time to order the world—there has been a growing trend toward individualism in society. This individualism is nothing more than the transfer of ideas from classical mechanics—the understanding of the world as a perfect and absolutely predictable machine—onto society and people. The world of industrialization, and even more so, the post-industrial world, are direct consequences of the physical laws that established the world as a void filled with isolated entities that produce by exerting forces on one another. In this world, freedom is increasingly understood as the preservation of each of these isolated entities—as each stone's preservation. Individual freedom is seen as isolation, as the absence of external forces that condition the individual. A stone is understood as free when it breaks free from the equilibrium held by the world around it and begins to fall and roll—although that could mean its destruction as it crashes into a stronger rock at the edge of the cliff.

The problem with this is that a world of isolated stones makes no sense. Stones only make sense when they form structures. Stones only have meaning when they establish relationships with each other. A brick alone is useless, but a group of bricks well placed together forms your home. Atoms of carbon, hydrogen, oxygen, nitrogen, phosphorus—by themselves, they're nothing but stones. But when ordered and related properly, they form the miracle of life. The stone that's in equilibrium as part of a cliff is part of something as beautiful and awe-inspiring as a cliff. The stone that breaks away to fall and roll—the most free it can be—is just another stone.

The universe is nothing more than a group of stones playing musical chairs, running as fast as they can so as not to be left without the chair that allows them to continue participating in the game. It's by being part of the game, not staying out of it, that stones can truly be free. If there's one thing quantum mechanics teaches us—regardless of how it's interpreted—it's that before playing the game, before establishing relationships with the rest, stones are nothing but potential. It's easy to confuse potential with freedom because it seems like the ability to be in more than one state at the same time—to be in superposition—but in reality, in practical terms, what potential means for the material world is that you're not yet anything. Is the stone that stays out of the game and can no longer participate—yet could potentially have sat in all the chairs at once—freer? Or is it the stone that momentarily gives up its potential, joins the game with the others, and can continue playing? Freedom lies in limits. Limits are the condition for the possibility of being truly free. A world that appears completely free is a chaotic world without established rules, where everything can be everything and yet nothing is meaningful. Imagine trying to play a game without rules—what kind of game is that? What freedom is there in that? There's not even a game, let alone freedom! The universe exists because it's a game with established rules (physical laws), because it has limits (the maximum speed of light in a vacuum), allowing you to break your symmetries and make at least part of your potential a reality. Only then does it give you the freedom to exist.

JAITZULF.

It's in that game, in that interaction, in the relationship formed between the stones, where everything takes on meaning. And if it's the relationship that gives meaning to things, maybe that's where the key to true freedom lies. Perhaps it's in defining the relationships between stones, not the stones themselves, that should capture our attention. After all, no one really cares about the stones.

# AND THE STONES.

JAITZULF.

*I'll take a shower*
*15:27*
*And if I have the energy, I'll go out for a bit*
*15:27*
*But right now, I only feel like changing the world*
*15:28*
*Moving Stones*
*15:28*
*14/04/2017*

# MEMORIES OF STONEHENGE.

Since I was a child, I've been fascinated by dolmens and other types of megalithic constructions. So much so that, as soon as I started working and had some money, the first thing I did was travel to visit Stonehenge.

In the Salisbury Plain, England, stands a circle of enormous, heavy vertical stones—Stonehenge. Its construction, dating back to around 3100 BC, is still a mystery due to the immense difficulty of moving such monoliths at that time. The purpose behind its construction is also unknown, although it's assumed to have been of great importance given the effort it must have taken to build. Theories about its purpose range from it being a site of sun worship—since the stones align with the solstices—to being a place of healing or a burial ground. Some suggest it might have served all these purposes at different times. Whatever its function, if it had one, it's clear that this monumental effort was made to create a community meeting point.

I visited the site during a particularly turbulent time in my life. After finishing my degree and a master's in physics, with my mind completely disconnected from the daily realities of life, adulthood was beginning to feel more like an obligation than an aspiration. All the ideas society had built for us millennials about what life should be like had crumbled, and the weight of all the burdens I hadn't shed along the way had become too heavy to bear. Like the stones of Stonehenge. Additionally, and probably the most significant factor in my state of mind, my grandmother had recently passed away. In that context, seeing those stones somehow helped to organize all that chaos. Those stones, built by who knows who, who knows when, and for reasons we can only speculate about, brought order to the confusing and chaotic thoughts swirling in my head, making them appear stable and comprehensible. Like stones always seem to do. On the bus back to London, I wrote the following:

"If there's one thing I've learned from studying physics, it's that everything boils down to symmetries and possibilities. Stonehenge exists because it broke some symmetry in trying to be symmetrical—and because there was a possibility for it to exist. Today, I don't know how many years later (I'm terrible with dates and don't feel like looking it up because it's irrelevant to the point), if you ask anyone about the possibility of carrying stones that size without advanced technology—or aliens—they'll tell you it's impossible. Impossible. But nothing could be further from the truth, of course. Humanity, somehow, has evolved into a state of impossibility, has grown comfortable and abandoned its role as a creator of possibilities. As if life had any other purpose. Let's break symmetries, create possibilities: let's move stones."

The next day, I cried as I walked through the streets of Southampton.

Reflecting again on this text, the story of Stonehenge, as seen from our perspective, is essentially the same story as that of the pyramids or countless other ancient constructions. They're there, we can observe them, we can even date them, yet we label them as impossible. Stories of impossibility, as is increasingly recognized, are often narratives designed more to discredit non-civilized worlds than to reflect reality accurately. It's obvious that none of these constructions were impossible to build at the time, because they were built. It's clear that the humans of that era were more capable than we make them seem now. That Stonehenge stands there, and we can't understand how or why— to the point where we resort to magical or science fiction explanations—says more about us as observers than about it. Once again, as with Schrödinger's cat, it's Stonehenge that defines us as observers, not the other way around.

# THE VOID.

The introduction of this book already puts forth the idea that the concept of the void is a mental construct in physics, one that has been used to build our entire understanding of the world. The evolution of physics, in practice, is the development of the content that fills that void. Starting with an empty space, an empty box to be filled, with nothingness as the foundation, we've developed a series of ideas to explain the world we inhabit. The world has been filled first with water, earth, air, and fire, then elements, matter, forces, energy, ether, electric fields, magnetic fields, quantum fields. The ether has been emptied, then replaced with quantum fluctuations; forces have been stripped away... Since physics entered social thought, the generalized view of the world is that of a place filled with things—us included—that follow a set of physical laws defined by equations, functioning like a precise, unstoppable, and insatiable machine. This image has wavered at times, requiring us to redefine certain concepts or accept a certain degree of inherent uncertainty in the world, but it has remained fundamentally rigid to this day, with new discoveries generally being adapted to fit this narrative, not the other way around.

Einstein's famous theory of relativity, so often referenced to imply that "everything is relative," is actually a colossal effort on his part to maintain this vision of the world as a singular machine that occupies space and whose underlying truth must be universal for all its components. General relativity admits that different observers may have different descriptions of the universe, but it never accepts that the truth itself can differ for different observers. That's why situations like the twin paradox become paradoxes: if the theory simply accepted that everything is relative, such paradoxes would never arise. The same applies to many of the problems posed by quantum mechanics. All of them —or many of them—stem from the insistence on maintaining that mental framework.

For centuries, we've been trying to move stones to fill our void, thinking it's the stones themselves, occupying physical space, that fill the emptiness. If I may be so bold, I believe it's time to face the reality that it's the relationship between us and the stone, the very act of moving the stone, that fills the void. Just as the quantum vacuum is filled with chaotic interactions that prevent it from ever being truly empty. Not because it contains matter or energy, but because there is constant, incessant interaction at the most fundamental level of the universe. It's the interactions that fill the void, not the particles. It's the act of moving the stones, not the stones themselves. Even if we begin with the void to understand the world, the conclusion is clear: it's the relationships that fill the world, that give meaning and significance to things. But I haven't written all these pages just to reach this conclusion. If the most accurate, precise, and scientifically validated theory to date —quantum physics—tells us that the void, as we understand it, doesn't exist—because even the emptiest existence possible isn't empty—then maybe we should stop starting with the void to understand and give meaning to the world. Even if it leads us to the same conclusion.

If we understand the universe as a collection of particles that are constantly interacting—exchanging energy, exchanging information—with one another, and if we understand that it's this interaction, this relationship between particles, that constructs the entire universe and all the meanings that our conscious minds process, then the most basic existence possible is the relationship between two things, not an empty box. So let's

JAITZULF.

replace the void with the relationship between two stones.

# MOVING STONES I.

It's been so long since I've lived with the idea that the void isn't empty that it has started to lose its meaning in my mind. I've normalized the idea so much that it's become part of the balance, part of my natural thought pattern. The concept has petrified in my mind. But it wasn't always like this.

The idea of the void and nothingness is closely tied to death. I must admit that, until I encountered this stone along the way, death had been a practically nonexistent topic in my head. Before I existed, there was nothing, and afterward, there would be nothing. That idea formed a perfect balance in my head: I didn't need to delve deeper into it. But once the idea that "nothing is nothing" was brought into my reality, I was no longer able to relegate it back to the realm of nonexistence. It has stayed with me, in an eternal present, interacting with all the other thoughts invading my mind, seeking to form a new balance that would allow it to become part of an order and finally rest in peace. To reach that point, I've had to go through a mental process that now seems impossible to me. Lifting that heavy stone, brought from so far away, and placing it in a way that brings balance with the rest of the equally heavy stones that shape my understanding of the world has been a task worthy of titans.

Throughout almost all of human history, we have tried to understand death. We've tried to give it meaning, as we do with everything. But no matter how hard we try, it remains an impossible task. Truly impossible, in this case. Religions and beliefs generated by our minds generally understand death in two ways: either as a continuation of life (whether through reincarnation or transitioning to another plane like heaven or hell) or as a passage into nothingness, into nonexistence, into being part of the void. What's curious is that — if we assume the second option as the real one — we all would have already been part of that nothingness, since we all have a birthdate and a clear beginning as beings — even though the components that make up our body and mind already existed in the universe beforehand, establishing other relationships that shaped other things. But none of us, absolutely none of us, can comprehend or imagine going back to being part of that nothingness. Once we become part of existence, nonexistence seems incomprehensible. Maybe that's why the only thing we can assume is the first case: that death is the continuation of life, even if it's in another plane, with different rules and perhaps other sensations, but still being something. The idea that the void isn't empty resonates with this: if absolute nothingness doesn't exist, how could it be possible to become part of it? Can you be part of something that doesn't exist?

# THE SEA.

Observing the sea is the closest thing to observing the universe from the outside. The sea is an entire universe unto itself. An infinite place (boundless in possibilities) constantly creating and destroying everything within it. A place that seeks to expand as far as possible, unable to stop, needing to generate disorder continually to allow new orders within itself. The sea is the sum of an infinite number of changing entities, and at the same time, it is only a single entity. Just like you are made up of an impressive number of cells united to be you, yet you are only one. When we observe the universe, it's as if that tiny cell that is part of our body were trying to observe the entirety of the body — an immensely difficult task to grasp from that position, even though from our own perspective, it's quite obvious to understand the boundaries of our body. The same applies to the sea. When we look toward the universe, we are fish trying to understand the sea.

Understanding our higher entity, understanding the unity we form with the rest of the universe, should only be possible if we could observe that entity from the outside. Only from outside things can we, in principle, comprehend a set as a unit. Yet despite that, despite how impossible it should be for us to understand ourselves as part of the universe's unity, for some reason unknown to us, we are an atypical part of its machinery that has acquired the ability to store information and use it to create and shape our own abstract spaces. And that abstraction allows us to construct narratives that, whether more or less accurate, at least lead us to the illusion that it's possible to create a universe-unit for ourselves.

In quantum mechanics, time is constantly accompanied by the unit i — the unit of the imaginary number plane — because the temporal evolution of the wave functions associated with quantum systems occurs in that imaginary plane, detached from reality, which is quantum superposition. Only when the system collapses through interaction with something do its properties become defined with real numerical values. If we take this literally, it tells us that time is clearly an artificial construct created by humans. A way of forcing our definitions — and their natural evolutions — beyond the moments when we interact with things. A construct meant to order everything before our eyes and avoid its natural tendency toward disorder. If I had to name the most unnatural thing in the world, it would undoubtedly be time. Nothing needs time to exist, except for human beings. Time is the embodiment of our capacity for abstraction. It's an illusion created by our ability to store information, by our memory. Memory, which is nothing more than the ability to eternalize what we have lived as if it were the present — the ability to interact eternally (as long as our memory lasts) with everything we have previously interacted with — is what generates the illusion that the universe operates on a temporal order. It creates the illusion of the cosmos, the illusion of apparent order. And with that, it allows us to create narratives, interpretations, explanations, and thus understand everything happening around us, while also compelling us to define everything through them. Time is by no means a defining magnitude of the universe, but it is the defining magnitude of our understanding of it, as part of it. Time is the only way we, from within the universe, can abstract ourselves from it and try to observe and understand it as a unit. It's our tool to unify the web of interactions that make up the universe, our tool to give it the entity of unity.

Time is the tool that allows us to bring order to the chaos we are a part of. It isn't the perception of cycles and repeated events that led us to construct the idea of time, as I mentioned earlier, but rather the opposite: it is the idea of time, our power of abstraction, of carrying the past with us, that allowed us to detect those cycles and repeated events, giving us the power to bring order to disorder. The power to construct a narrative of the world that opposes its most basic principle: the tendency toward disorder. Time opened the door to knowledge. Time is the relationship between our memory and the world.

# MOVING STONES II.

Observing yourself from the outside, understanding yourself as a being, as the sum of all the cells that compose you—wow, that is a heavy and difficult stone to move.

Self-perception as a being, creating an identity that one feels comfortable with and that is also perceived by others in the same way, is a titanic process for any human. Unless one settles for accepting the identity others have assigned to them, of course. It's also a process that inevitably depends on one's relationship with the environment. It's only through countless relationships, by exposing oneself to various situations, that this becomes possible. Only then can we see ourselves from the outside. And of course, by using our power of abstraction—time.

Creating identity is writing our own narrative. Identity, understood as an absolute definition of a being, inevitably requires time, a product of our capacity to store information. However, this doesn't align with how I live or understand it. Identity doesn't require time. Identity isn't absolute; it's relative. It's the way you relate to your environment, to other beings, objects, etc., that essentially defines you, not the narrative that is created from it. And as a category that doesn't depend on time, it means identity doesn't persist over time. Identity is a momentary fact, quantized in finite leaps. It isn't a continuum we carry with us through time. Instead, just like an electron in superposition, we are only defined when we relate to something else, and then we return immediately to indefinition.

## A STORY ABOUT THE VOID.

Let's be honest and clear: no human being ever, at any time, needs self-perception or self-definition for themselves. It's always for the environment, always in relation to others. We define ourselves for others so they can place us on their timeline, so they can predict how to interact with us. The abstraction and attempts to see ourselves from the outside to understand ourselves as a unit are a social consequence of our inability to understand ourselves as part of a collective whole. They result from the need we've created to organize relationships before they happen to understand them—just like time is a mental construct we've created to overcome the challenge of understanding ourselves as part of the whole universe.

Understanding this—I'll tell you as I finally place this stone beside the first one I brought—means embracing the idea that since it is the relationship between things that defines and gives them identity, trying to define oneself is as absurd as asking what our relationship with nothingness is. Self-perception is a consequence of time, an illusion generated by our memory as it allows us to interact constantly with our past, creating a narrative of our existence. Self-perception is our relationship with the collage of past experiences we store in our memory, just as our perception of the world is the relationship with its collage of past events while we remain slaves to time. Nothingness, meanwhile, cannot relate and therefore is not a real physical entity. Just as your self-perception or conception of the world won't be real as long as we remain bound by time.

# PHYSICS.

The claim that the study of physics is humanity's greatest achievement is irrefutable. Studying nature, the hostile environment in which we live, has allowed us as a species to dominate the planet we inhabit in an absolute, indisputable way. Exploring the world, interacting with it, exploiting and understanding the possibilities it offers us, is undoubtedly the collective task that defines us as a species. In fact, it's the only collective task possible. Everything is physics. Everything we do, every action a human takes, is the study of physics. To exist is to interact with the environment, to relate to it. And from the moment every action we take is processed and stored in our memory, everything becomes a study—a study of our physical world.

Physics is often presented as this elitist science belonging to brilliant minds, to geniuses who can read the magical formulas written by nature. And that's a terrible mistake. What these men, whose surnames I've mentioned numerous times in this text, did was nothing more than relate the information they obtained from interacting with the world—something we all do constantly and at every moment. Translating those relationships into a universal language—the mathematical language—that allows us to control, quantify, and use those relationships to our advantage is undoubtedly an exceptional achievement. But it isn't a task exclusive to special people; it's simply a task available to those privileged enough to dedicate their lives and efforts to worrying about external matters because their own basic needs are already covered. Newton wasn't a genius; he was a boy whose life was already sorted. Einstein wasn't a genius; he was a man with a woman enslaved. Only those who are privileged to live comfortably have the capacity to think about everything else. And because of this, everything these people have deduced from the relationships they observed is inevitably marked by bias. I want to be crystal clear: I'm referring to the deductions, not the relationships themselves. The relationships observed are indisputable. Any of the equations accepted in physics can be perfectly tested by experiments, no matter who conducts them or where. There's no doubt they are accurate relationships. The problem lies in their interpretations, in the narratives built around them. That's where bias enters. Pre-existing beliefs, ideas assumed to be real by our minds, the foundations of how we relate to the world—these influence our interpretation of it, inevitably. And if all—or most—of physics has been interpreted by men of a specific, clearly defined profile, then the view of the world they are leaving for the rest of us is also shaped by that. Despite all scientific advances, despite everything that's been discovered over recent years, the image of the world constructed by physics remains almost unchanged. The perception of the world people have when hearing the word "physics" is always the same: machinery, cold, calculating, sterile, almost robotic. They see the world as a soulless machine, a computer simulation. As a collection of isolated things or data, perfectly defined, manageable, and quantifiable. It's horrifying—hellish.

Is this the only possible interpretation of physics, of our world? The truth is that the more you read about physics, the more you study the relationships that define the world, and the deeper you dive into the essence of what these equations dictate, the harder it is to sustain this image. Quantum mechanics is to this worldview what the spherical Earth was to the Catholic Church: it should have signaled the end of this view of the world, yet all that was done was first deny it, and then attempt to merge both perspectives into a middle ground that inevitably leads to contradiction. Religion clings to its most fundamental pillar—God. Who is the equivalent of God in the dominant interpretation of physics? I'd say it's time.

If Nietzsche wrongly announced the death of God, I invite you to kill time—metaphorically and literally.

Time, as I've been developing throughout, is the relationship we've generated with the world to bring order to it. All the order we've established, scientifically and socially, requires that we first construct the concept of time. Without time, we wouldn't have equations describing the evolution of physical systems, no technology, no ability to predict and shape the behavior of everything around us. We wouldn't have history, nor would we have the need to create motivations for the future. Without time, it's impossible to obsess over productivity. Without time, it's impossible to organize society. Time is what allows us to anticipate the world, so it ceases to be a hostile place for us. Time is God.

The issue with time—which, I insist, is the most powerful concept we've developed for our benefit—comes precisely from its immense power. The power to order chaos is undeniably the greatest power in the universe, but like any power, it isn't without consequences. Ordering chaos means losing its nuances, defining the undefined—and often indefinable. Establishing order requires first setting a framework of thought where everything we want to describe must be defined. In the case of time, it requires not just a precise definition of things at a moment, but a definition across time. The construction we've made of time demands that everything around us is defined in the past, present, and future because only then can it be understood within that framework. Everything around us must follow a life line—defined before us and continuing after. To some degree, in whatever form, but defined. Time forces all existence to have a defined past, present, and future and to connect these three.

Classic physics, with its absolute nature, already shows how, due to its precision, it can easily lead back to chaos (butterfly effect). Though modern physics has gradually shed this absolute need, opening up to understanding time as relative (general relativity) and even abandoning its role as a parameter (quantum physics), socially, we have almost no choice but to pretend that these effects don't apply to us and maintain the rigid notion of absolute time governing our lives. Because giving up order is terrifying—naturally. But perhaps it's time to open that drawer and never close it again. On reflection, everything that frightens us about general relativity and quantum physics is simply the fact that this established order, this absolute time, isn't real. Quantum physics, the most terrifying reality Western thought has faced, is simply proof that this ultra-definition, this established order, is an illusion we've imposed on the world. What quantum physics shows us is that assuming everything is defined along its timeline is a grave mistake. Nature isn't defined by a past, present, and future. Nature is defined only in each interaction, in each moment. It is defined through relationships. Particles define themselves based on the relationships they establish in each moment with others around them. Quantum superposition is, in reality, simply accepting that the electron lives in the now —it doesn't have a past. It exists only in the moment it interacts with something else. The electron doesn't have strange behaviors like wave-particle duality or being in multiple states waiting for us. The electron doesn't have a past. It's just an electron when we interact with it, and it defines itself only in that moment based on our relationship with it. Our obsession with forcing the electron to have a defined past is what makes superposition seem strange. If something persists over time, it's only because it maintains a more stable relationship. And it's only because we have the capacity to store the past in our minds that, when faced with a lasting relationship, we get the feeling that things are defined at all times.

Time is, in essence, what separates humans from everything else that exists. But. But it's also our greatest burden.

Time forces us to understand the world based on defined terms and control everything, including ourselves. It is, at once, the most liberating tool in terms of shaping our world, and the heaviest burden we carry. This need to define everything and control it in temporal coordinates is also the root of most major human conflicts: wars, hatred of others, incomprehension between people who don't meet the expectations generated by the definitions we've created, and many mental health problems like anxiety, depression, and stress. And that's why I propose we kill time—or at least take steps to blur the cage it has become for us. In this sense, embracing the concept of time as it's understood in quantum physics could be the first step: embracing the idea that time should only be a useful parameter to describe the evolution of a system, without the need for absolute definition across its entire history. Embracing the idea that things are defined in each second, in each moment, through their relationship with their environment, rather than being something permanent throughout their existence, is a first step in understanding that it's actually relationships that define everything, not things themselves.

Killing time means understanding that Stonehenge exists only from the moment our ancestors gave those stones a relationship by placing them together. The story of how our ancestors moved those stones doesn't belong to Stonehenge but perhaps to our ancestors and each of the stones themselves. Killing time is accepting that the universe exists only after the Big Bang, once it began to disorder itself and establish relationships between its components. Killing time is accepting that the story of how that state before was reached, the story that might precede the formation of our universe, doesn't belong to us. Just as the story of everything that happened afterward doesn't belong to the photon that emerged and continues wandering through the void without interacting with anything. Killing time is realizing that our existence, limited as it is, is just another event in this sea of possibilities. The elements that make us formed one order before and will form another order afterward, with a different meaning. We are a bunch of stones that, for some reason, have renounced their apparent freedom and have decided to organize themselves to allow us to participate in the game of existence—the only true freedom. When death comes for us, it will simply be the end of just another event in the universe.

# MOVING STONES III.

*Killing time.* Honestly, I don't know how I'll be able to move this stone.

The progress of human society, paced by the passage of time, brought with it the understanding of humans, those who form societies, as individuals. It brought the need to define those components. Just as we needed to define fixed behaviors and properties for the stones that form the universe to describe their evolution over time, we did the same with ourselves. This categorization and definition of individuals within society follows exactly the same motivation as defining stones: to control the individuals that form the collective and to have an orderly reading that prevents chaos. In that sense, its usefulness in establishing peaceful, structured societies in the world is undeniable. But it's not our truth, not the truth of the world. And when this society—so well-versed in the necessity of that categorization for the survival of the order that protects us—encounters an individual who doesn't fit within these categories, the misunderstanding, hatred, and attack toward that person are palpable, to say the least. And it's understandable, after all, they are seen as an attack on the established order. They are a door to disorder, to chaos, to killing time.
If having an external self-perception of our identity as individuals is already difficult—if not impossible—in this context described above, it becomes utterly impossible for the vast majority of society. I'm not just talking about queer people, like when I spoke

about Schrödinger's cat, but about almost anyone who is part of society. Even if we try to see ourselves as simple stones, we are polyhedral stones with many facets, many irregularities, with almost all our symmetries broken. We are made up of various minerals in unmeasurable proportions. We are stones formed from stardust, capable of moving, in different ways, with abilities as incredible as abstracting and creating other planes of existence unique to us. It's possible to categorize us if we want to talk about specific things at specific times and only analyze a particular defining part. But not to categorize us entirely as complete beings, let alone across time. Pretending that our skin color, place of origin, sex, or preferences fully categorize us is unnecessary and ridiculous. More absurd still when this is done even before our time begins to run. Without being born yet, without having interacted with anything or anyone, and knowing nothing more than the place of birth, genitalia, or the skin color of our parents, we insist time and again on projecting the existence of future humans. We are obsessed with organizing others in time. And it's understandable because that's the only way we've learned to understand ourselves. But, because of this, we're not only causing immense suffering to ourselves by forcing us to fulfill a role on that tightly defined, narrow timeline, walking a tightrope and constantly falling, injuring, and destroying ourselves, but we are also missing out on a vast range of incredible possibilities and relationships.

To move this stone, to kill time and free ourselves from the need to follow order without falling into absolute chaos, we must only understand that while time is an abstraction tool, useful and functional, it's not the truth that should govern us. In other words, just as quantum physics uses time for its benefit but opens up a world defined by constant relationships that exist beyond the limits of time, we should try to see the world from this new mental framework. A world in which every definition is open to a vast potential—that allows magical-seeming things to happen—

and is realized in each moment through the relationship between its parts. It's not only possible; it's the world we already live in. Even if it scares us to acknowledge it, even if it terrifies us to disorder the order we've already established.

I'm afraid this stone is so, so big that it's impossible for me to move it alone. As I try to move it, memories from my childhood flood my mind—voices calling me "fag" just because I didn't fit into the roles society had defined for me as a man. I remember feeling trapped by society for more than half of my life due to the expectations they had built for me based on absurd things like having a penis, having more female friends than male ones during childhood, not liking football, dressing up as a fairy, or painting my nails. I take a deep breath, straining to hold the stone in my arms. I also think about how society categorizes you and forces you, if you don't fit the role, to at least assume the narrative they have prepared for you. Society begins to accept that you're different but insists on not only telling you that you're different but also how you should feel about it. I think about how the burden of guilt and responsibility is placed on us—the responsibility for others' failure to understand us. As if there were something to understand. The stone grows heavier in my hands. I recall the screams of that idiot at Róisín Dubh, yelling that I was disgusting and should be dead for kissing another man. The stone wobbles in my grip. I think about how adolescents today still turn to black clothing to avoid showing any hint of deviance from the mold imposed on them. I think about how people still believe that putting a unicorn shirt on a boy is "dressing him up." My strength wanes, but I make one last effort to hold the stone. I think, in general, about how even today I hear people speak of non-normative individuals as strange beings, as if they are the ones causing conflict simply for freeing themselves from the yoke of understanding the world through the Copenhagen interpretation, for refusing to give power to the observer, for wanting to explore their potential, for wanting to live their own Big Bang, create their own universe, live their own time, their own order, and generate disorder along the way, loving whoever and however they want. As if life could have any other purpose. The stone slips from my grasp and falls to the ground.

I'm sorry.

# LOVE.

To love is to live. If existence is a network of energy exchanges, a network of interactions, of relationships between things, constant and insatiable, if existence is to disorder the world, then to live is to have the ability to make those changes against the established order. To live is the ability to move stones at our will. What does not live, what changes and transforms, only does so through the impulses and limits marked by physical laws—in a completely involuntary, mechanical way, cold and calculated. Inert matter is indeed mechanistic in the classical sense of physics: it has no choice but to move toward equilibrium; its destiny is entirely determined. However, unlike what classical physics dictates, it does not do so in isolation. Stones are not alone in the world. Stones constantly interact with everything within their reach. And for reasons unknown to God, part of that environment gained the ability to live a long time ago—the ability to transform the world for its benefit, to disorder it in order to create its own order. To live is to love stones; to live is the ability to give your energy to others. It's to be a stone that can move other stones at will (or need). It's to be a stone that moves itself as it needs, a stone entirely asymmetrical, one that breaks with the strongest, most binding symmetry in the world: freedom.

One of my favorite quotes by Carlo Rovelli is that "the world is not made of stones, but of events like kisses." You can ask where a stone will be tomorrow, but it makes no sense to ask where a kiss will go. A kiss happens, exists for a limited time, it's a unique event.

And that's how I understand the world—as a network of kisses. A network of events. A network of relationships. The insistence on interpreting physics as a network of stones crashing into each other is a task that long ago lost its purpose. The world we live in offers far more than any machine—it is much more than a computer simulation. It isn't a cold, calculated place that's perfectly measurable and rational. It's a place filled with potential, possibilities; it's a place full of magical, unique events. It's a wild party where, no matter how much you try to control it, you never know how it will end. The world is defined by journeys, by paths, by the relationship between the beginning and the end of things. The world is a passionate and exciting place.

The magic of the world is so immense on its own that it doesn't need us to keep relegating things to the drawer of magical phenomena. All phenomena in the world are magical. Every phenomenon is as important as the first kiss. Quantum superposition, the twin paradox, or quantum entanglement are only especially magical to our eyes because we insist on reading the world in the wrong way. All these phenomena are just consequences of ignoring the relational nature of existence. All these phenomena can be interpreted simply by focusing on defining the relationship between the parts involved and not each part. And that doesn't make the world any less magical. On the contrary, the fact that these phenomena can be explained more simply by embracing the relational nature of things makes everything even more magical. It makes everything, in itself, a magical event. The world is much more beautiful when viewed from the perspective of someone who understands that everything they perceive, everything that happens, everything that has happened, is happening, and will happen in this existence is an event. An event as special as exchanging virtual bosons at a distance, exchanging energy, loving. To understand that the entire universe is itself a great event, the result of a chain of interactions between its components, is far more magical than any other fictitious phenomenon.

The universe is a whole that, for some reason, broke into parts that have done nothing but try to reach each other ever since. They have never stopped sending signals—kissing each other—saying, "I'm here, you're not alone."

# MOVING STONES V.

To understand that quantum physics tells us that existence is only understood through interaction with the rest of the world around us shouldn't surprise us so much: if you're standing on the ground, it's because the ground holds you up; if you're alive, it's because you're breathing the air around you; if you speak, it's because that same air vibrates; if you came into this world, it's because your parents interacted... our existence has always depended on the interaction with and of the environment that surrounds us. And if our existence has always been conditioned by our environment, why all the fuss, then, about understanding that things are only defined insofar as they interact with something? Why so much fear in accepting what quantum physics dictates? Measuring, feeling, living, existing has always been about interacting, about establishing relationships. Quantum ideas are often used to blow minds with their counterintuitive concepts, but are they really? Could it just be that we've been deceived for so long that ideas that are actually the most natural seem crazy to us now? I mean, it wasn't too long ago that it seemed crazy for a woman to vote or to freely kiss her friend, and arguments were made about what was "unnatural."

The idea that the world—reality—exists outside of interaction—observation—between at least two parts, I'm going to be bold and say, is what's truly unnatural. Unless you believe in God or abstract yourself in time. The common external reality, spacetime as an entity that exists before us and contains us, is a reality that only a god could experience. Only a god would have the capacity to interact with everything at every moment to know its state. And your perception that reality works that way is likely because, no matter what Nietzsche said, God never died—we just turned him into a clock. But if you think about it for even a moment, if you reflect on your daily experience, it's obvious that the simple fact that you are in one place or another, at one time or another, changes the reality of the world. The observer has always interfered with the world, as inseparable a part of it as they are, no matter how hard they've tried to make you believe otherwise. The thing is, they don't interfere because they observe—they interfere simply by existing, by being, by being present. To exist is to interact with the environment, ergo, to exist and to interfere with reality are synonymous. Does that mean that the electron in superposition before it interacts with anything doesn't exist? I don't know, but even if it did, if it didn't interact with anything, it might as well not.

In any case, I bring with me one last stone while writing this, one that resolves any conflict here. It doesn't matter if the electron exists or not. Physically speaking—again, physically—the only thing that is defined is the relationships. The only certainty is the interaction between that stone we've called an electron and the stone we call the measuring instrument. Understanding this, accepting this, destroying the wall in our minds that forces us to define objects, beings, individuals, and the stage they inhabit (spacetime) and instead defining the relationships is a giant leap toward gaining degrees of freedom and embracing the infinite possibilities the world offers us. Earlier, I said that understanding the world this way resolves the great conflicts posed by modern physics. Let's see why before I place this last stone.

In the case of quantum superposition, I think by now it's understood why, since if there's no need to define the parts, but rather the relationship that occurs between the parts in a physical phenomenon, this concept falls away by itself: we call the state of superposition the range of possible relationships we can establish with the object we have yet to interact with.

Quantum entanglement? It's only a conflict because, once again, we believe that the properties we measure belong to the objects of study and not to the relationship we establish with them. In this experiment, we have two particles that have established a prior relationship—thus defining each other for each other—we separate them and measure them later, separately, defining their relationship to us. Understanding that the relationships between the parts that interact are what's defined, and not the parts themselves, removes the conflict and "spooky action" so often associated with quantum entanglement. In this interpretation of existence, each part involved defines itself for the other at the moment they interact—both particles define themselves in relation to the observer when the observer observes them and not before.

The same applies to the twin paradox, even though it belongs to a different branch of physics. And it's actually how this paradox is commonly resolved: by accepting that each twin's relationship with the world is different, giving them their own different times, and reconciling into a single possible reality when the two meet again and discover that time has passed differently for each.

In the end, to reconcile the world of large things and the world of small things, we may just need to accept that there are neither large nor small things—just things that are large or small in relation to each other, as has always been the case. But accepting this, accepting the absolute relativity of all reality, appears terrifying—at least at first. Does this mean there is no truth? Does this mean that, contrary to what Einstein said when establishing the principles of general relativity, the truth is relative to each of us, and there is no underlying, reconciling truth? How do you order a world defined by relationships, infinite relationships that

JAITZULF.

occur incessantly and fleetingly? Is reality fleeting? Does nothing last beyond our memory? So then, what? I'm afraid that in this view of the world, the only truth we're left with is kisses; the only truth left to us is love.

# THE STONES.

Now that I have these three giant stones with me, let me arrange them into a beautiful dolmen to honor the death of time, to honor the void, to honor the freedom of chaos and how its disruption allows us to enjoy this wonderful game we call life.

ΔΔΔ

JAITZULF.

ΔΔΔ

Allow me to stay here a while, observing it and simply enjoying its existence, without any other pretense. After all, they are just three stones whose apparent order I try to give meaning to. After all, it's only a matter of time before they become disordered again. I hope, by then, I'll have the strength to move the fourth stone I left behind.

ΔΔΔ

A STORY ABOUT THE VOID.

ΔΔΔ

JAITZULF.

**To be continued.**

**Memories of *Una historia sobre el vacío*.**

As I was finishing this book, the narrative I had originally envisioned for it started to blur before my eyes, disorganizing and reorganizing itself into a multitude of entirely different ideas. I suppose there's no fighting the one inevitable truth: the chaos of the void. That other story, I fear, needs a different medium than the void to sustain itself. That other story, I fear, requires different relationships in order to be established.

To Yaiza, for being my eternal present.
To Ainhoa, for being the first to disrupt my order.
To Borja, for teaching me to see the sea from Gorrondatxe.
To all who have taught me physics.
To Thiago, for being the first to disrupt my order.
To everyone who has disrupted my order afterward.
And to those who have tried to restore it.
To Ane Gálvez, for sharing my chaos.
To Ane Ariño, for allowing me to live in superposition, even as we interact.
To all my students, for reminding me every day of the importance of disrupting the world.
To all the people silenced by time.
To Salud, whose light is the only eternal event this existence will ever witness.
To Qasem, for taking care of me when I needed it most.
To Antonio, for teaching me the importance of moving stones.
To Jon, for reminding me of the importance of connection.
To Keir and Fur, for reminding me every day that I am not a stone when I was about to turn into one.
To mom and dad, for gifting me existence.
~ Jaitzul, 2024. ~

www.ingramcontent.com/pod-product-compliance
Lightning Source LLC
Chambersburg PA
CBHW071054240526
45471CB00015B/1873